Annual Energy Outlook 2017

with projections to 2050

January 5, 2017
www.eia.gov/aeo

Table of contents

Overview/key takeaways

EIA's Annual Energy Outlook provides modeled projections of domestic energy markets through 2050, and includes cases with different assumptions of macroeconomic growth, world oil prices, technological progress, and energy policies. With strong domestic production and relatively flat demand, the United States becomes a net energy exporter over the projection period in most cases.

The Annual Energy Outlook provides long-term energy projections for the United States

- Projections in the *Annual Energy Outlook 2017* (AEO2017) are not predictions of what will happen, but rather modeled projections of what may happen given certain assumptions and methodologies.

- The AEO is developed using the National Energy Modeling System (NEMS), an integrated model that aims to capture various interactions of economic changes and energy supply, demand, and prices.

- Energy market projections are subject to much uncertainty, as many of the events that shape energy markets and future developments in technologies, demographics, and resources cannot be foreseen with certainty.

- More information about the assumptions used in developing these projections is available shortly after the release of each AEO.

- The AEO is published pursuant to the Department of Energy Organization Act of 1977, which requires the U.S. Energy Information Administration (EIA) Administrator to prepare annual reports on trends and projections for energy use and supply.

What is the Reference case?

- The Reference case projection assumes trend improvement in known technologies, along with a view of economic and demographic trends reflecting the current central views of leading economic forecasters and demographers.

- It generally assumes that current laws and regulations affecting the energy sector, including sunset dates for laws that have them, are unchanged throughout the projection period.

- The potential impacts of proposed legislation, regulations, or standards are not reflected in the Reference case.

- EIA addresses the uncertainty inherent in energy projections by developing side cases with different assumptions of macroeconomic growth, world oil prices, technological progress, and energy policies.

- Projections in the AEO should be interpreted with a clear understanding of the assumptions that inform them and the limitations inherent in any modeling effort.

What are the side cases?

- Oil prices are driven by global market balances that are mainly influenced by factors external to the NEMS model. In the High Oil Price case, the price of Brent crude in 2016 dollars reaches $226 per barrel (b) by 2040, compared to $109/b in the Reference case and $43/b in the Low Oil Price case.

- In the High Oil and Gas Resource and Technology case, lower costs and higher resource availability than in the Reference case allow for higher production at lower prices. In the Low Oil and Gas Resource and Technology case, more pessimistic assumptions about resources and costs are applied.

- The effects of economic assumptions on energy consumption are addressed in the High and Low Economic Growth cases, which assume compound annual growth rates for U.S. gross domestic product of 2.6% and 1.6%, respectively, from 2016–40, compared with 2.2% annual growth in the Reference case.

- A case assuming that the Clean Power Plan (CPP) is not implemented can be compared with the Reference case to show how the absence of that policy could affect energy markets and emissions.

Energy consumption varies minimally across all AEO cases—

Total energy consumption
quadrillion British thermal units

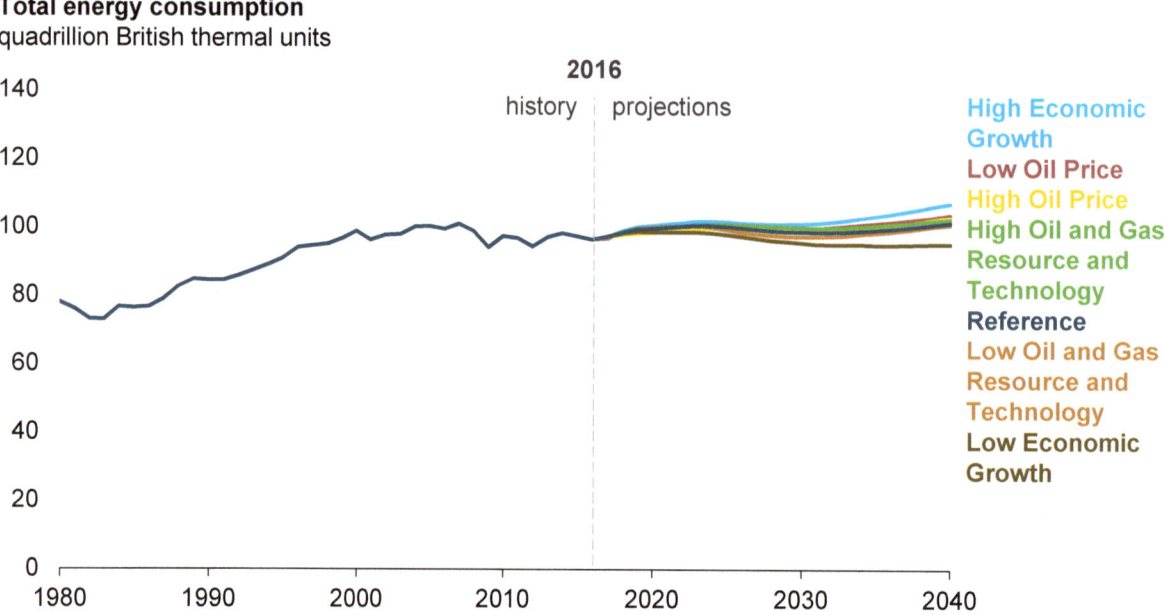

—bounded by the High and Low Economic Growth cases

- In the Reference case, total energy consumption increases by 5% between 2016 and 2040.

- Because a significant portion of energy consumption is related to economic activity, energy consumption is projected to increase by approximately 11% in the High Economic Growth case and to remain nearly flat in the Low Economic Growth case.

- Although the Oil and Gas Resource and Technology cases affect the production of energy, the impact on domestic energy consumption is less significant.

- In all AEO cases, the electric power sector remains the largest consumer of primary energy.

- Projections of total energy consumption (and supply) are sensitive to the conversions used to represent the primary energy content of noncombustible energy resources. AEO2017 uses fossil-equivalence to represent the energy content of renewable fuels.

Domestic energy consumption remains relatively flat in the Reference case—

Energy consumption (Reference case)
quadrillion British thermal units

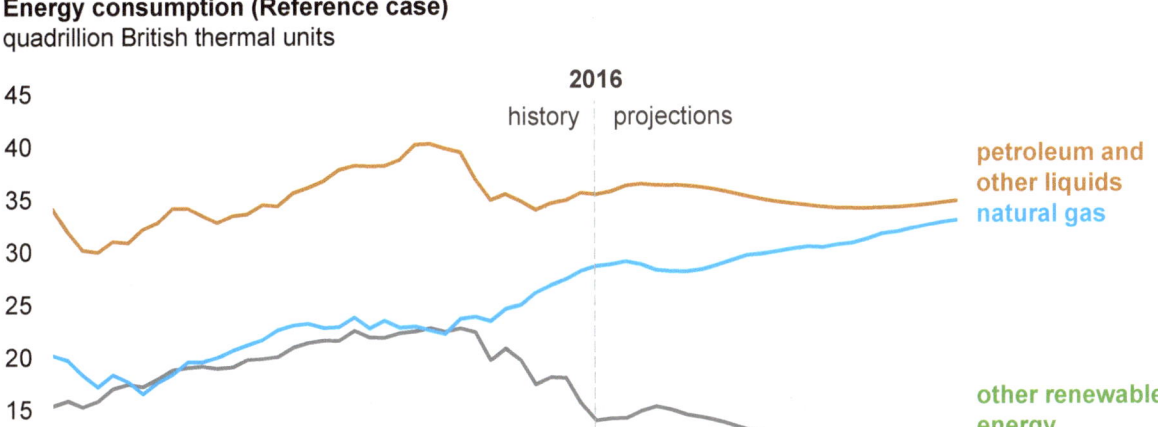

—but the fuel mix changes significantly

- Overall U.S. energy consumption remains relatively flat in the Reference case, rising 5% from the 2016 level by 2040 and somewhat close to its previous peak. Varying assumptions about economic growth rates or energy prices considered in the AEO2017 side cases affect projected consumption.

- Natural gas use increases more than other fuel sources in terms of quantity of energy consumed, led by demand from the industrial and electric power sectors.

- Petroleum consumption remains relatively flat as increases in energy efficiency offset growth in the transportation and industrial activity measures.

- Coal consumption decreases as coal loses market share to natural gas and renewable generation in the electric power sector.

- On a percentage basis, renewable energy grows the fastest because capital costs fall with increased penetration and because current state and federal policies encourage its use.

- Liquid biofuels growth is constrained by relatively flat transportation energy use and blending limitations.

Energy production ranges from nearly flat in the Low Oil and Gas Resource and Technology case—

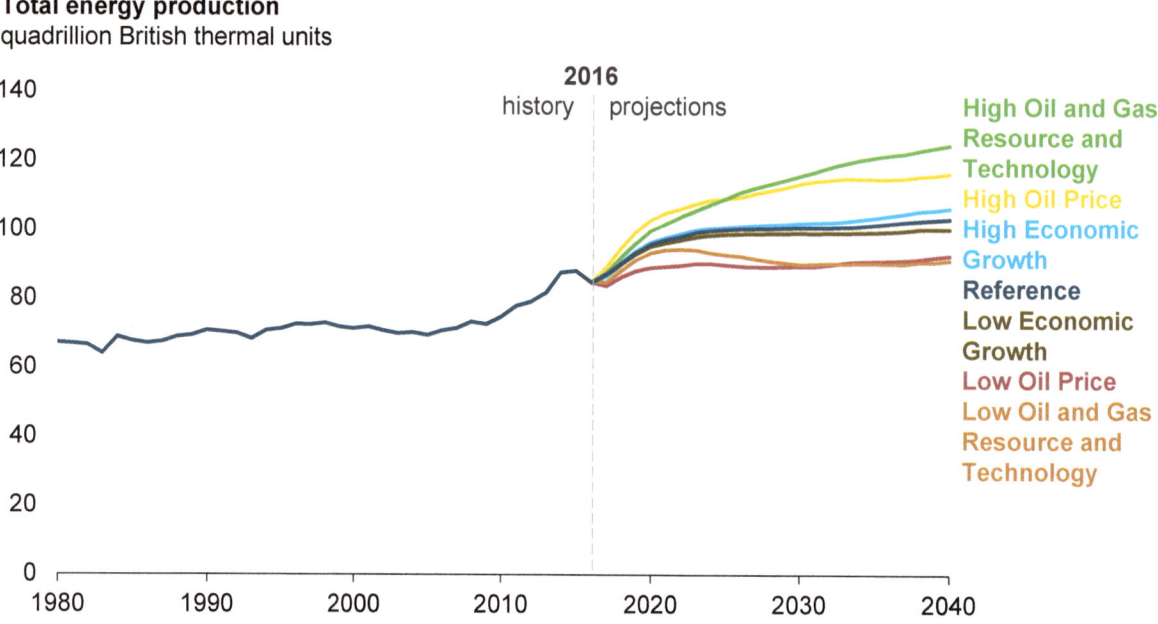

Total energy production
quadrillion British thermal units

—to continued growth in the High Resource and Technology case

- Unlike energy consumption, which varies less across AEO2017 cases, projections of energy production vary widely.

- Total energy production increases by more than 20% from 2016 through 2040 in the Reference case, led by increases in renewables, natural gas, and crude oil production.

- Production growth is dependent on technology, resources, and market conditions.

- The High Oil and Gas Resource and Technology case assumes higher estimates of unproved Alaska resources; offshore Lower 48 resources; and onshore Lower 48 tight oil, tight gas, and shale gas resources than in the Reference case. This case also assumes lower costs of producing these resources. The Low Oil and Gas Resource and Technology case assumes the opposite.

- The High Oil Price case illustrates the impact of higher world demand for petroleum products, lower Organization of the Petroleum Exporting Countries (OPEC) upstream investment, and higher non-OPEC exploration and development costs. The Low Oil Price case assumes the opposite.

U.S. energy production continues to increase in the Reference case—

Energy production (Reference case)
quadrillion British thermal units

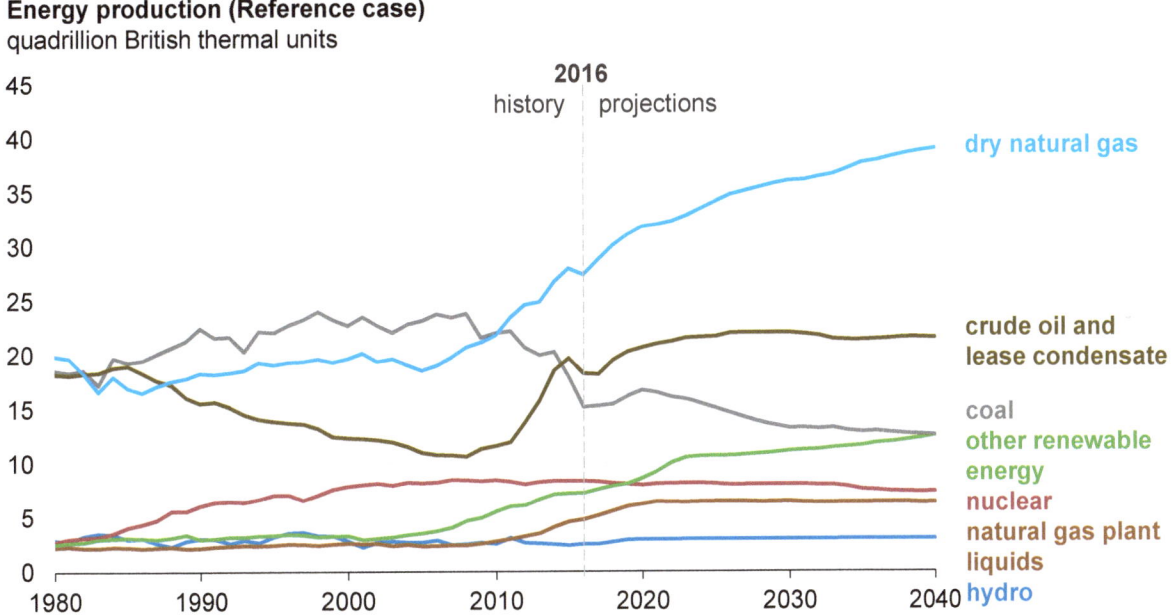

—led by growth in natural gas and renewables

- Natural gas production accounts for nearly 40% of U.S. energy production by 2040 in the Reference case. Varying assumptions about resources, technology, and prices in alternative cases significantly affect the projection for U.S. production.

- Crude oil production in the Reference case increases from current levels, then levels off around 2025 as tight oil development moves into less productive areas. Like natural gas, projected crude oil production varies considerably with assumptions about resources and technology.

- Coal production trends in the Reference case reflect the domestic regulatory environment, including the implementation of the Clean Power Plan, and export market constraints.

- Nonhydroelectric renewable energy production grows, reflecting cost reductions and existing policies at the federal and state level that promote the use of wind and solar energy.

- Nuclear generation declines modestly over 2017–40 in the Reference case as new builds already being developed and plant uprates nearly offset retirements. The decline in nuclear generation accelerates beyond 2040 as a significant share of existing plants is assumed to be retired at age 60.

The United States becomes a net energy exporter in most cases—

Net energy trade
quadrillion British thermal units

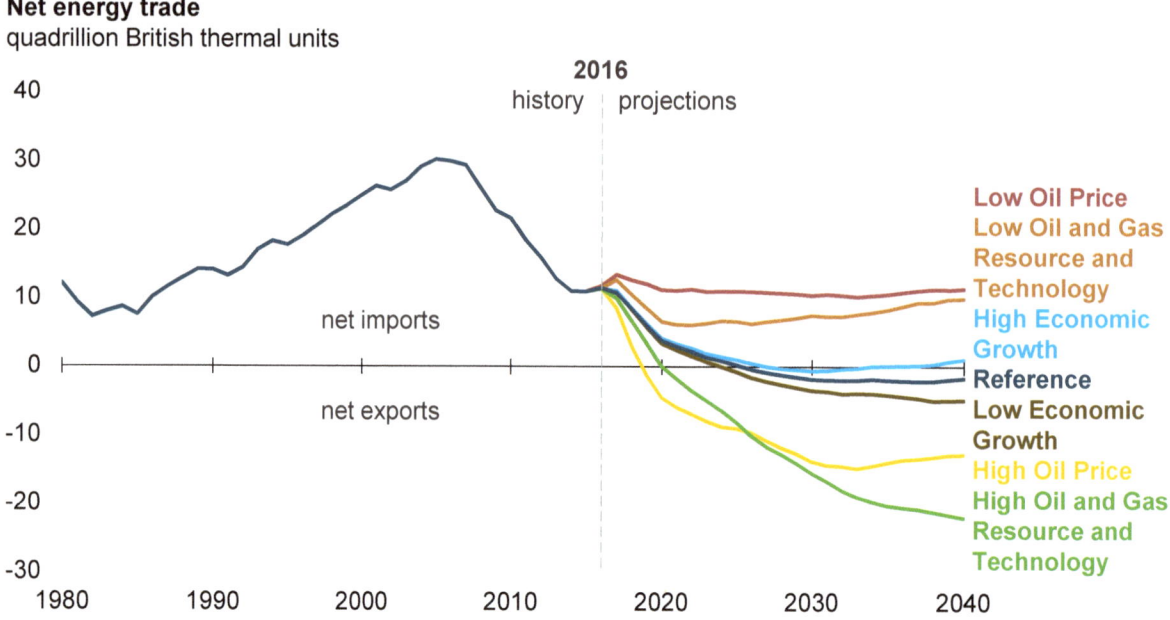

—and under high resource and technology assumptions, net exports are significantly higher than in the Reference case

- The United States is projected to become a net energy exporter by 2026 in the Reference case projections, but the transition occurs earlier in three of the AEO2017 side cases.

- Net exports are highest in the High Oil and Gas Resource and Technology case as favorable geology and technological developments combine to produce oil and natural gas at lower prices.

- The High Oil Price case includes favorable economic conditions for producers, but consumption is lower in response to higher prices. Without substantial improvements in technology and more favorable resource availability, U.S. energy production declines in the 2030s.

- In the Low Oil Price and Low Oil and Gas Resource and Technology cases, the United States remains a net importer over the analysis period.

- In the Low Oil and Gas Resource and Technology case, the conditions are unfavorable for U.S. crude oil production at levels that support exports.

- In the Low Oil Price case, prices are too low to provide a strong incentive for high U.S. production.

The United States becomes a net energy exporter in the Reference case—

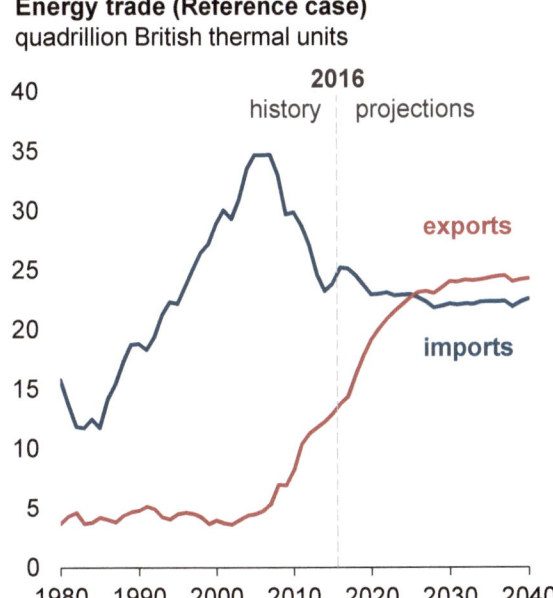

Energy trade (Reference case)
quadrillion British thermal units

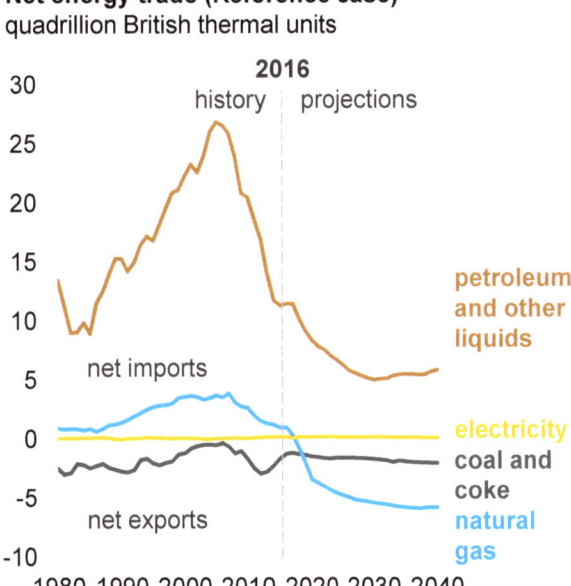

Net energy trade (Reference case)
quadrillion British thermal units

—as natural gas exports increase and net petroleum imports decrease

- The United States has been a net energy importer since 1953, but declining energy imports and growing energy exports make the United States a net energy exporter by 2026 in the Reference case projection.

- Crude oil and petroleum products dominate U.S. energy trade. The United States is both an importer and exporter of petroleum liquids, importing mostly crude oil and exporting mostly petroleum products such as gasoline and diesel throughout the Reference case projection.

- Natural gas trade, which has historically been mostly shipments by pipeline from Canada and to Mexico, is projected to be increasingly dominated by liquefied natural gas exports to more distant destinations.

- The United States continues to be a net exporter of coal (including coal coke), but its exports growth is not expected to increase significantly because of competition from other global suppliers closer to major markets.

Energy-related carbon dioxide emissions decline in most AEO cases—

Energy-related carbon dioxide emissions
billion metric tons of carbon dioxide

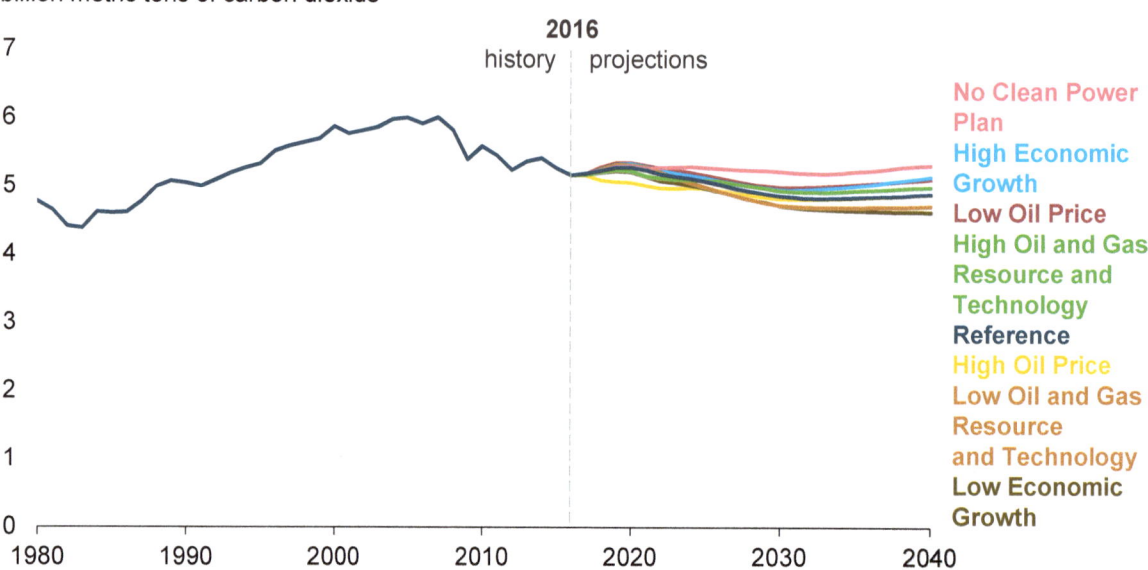

—with the highest emissions projected in the No Clean Power Plan case

- The electric power sector accounted for about 40% of the U.S. total energy-related carbon dioxide (CO2) emissions in 2011, with a declining share in recent years.

- The Clean Power Plan (CPP), which is currently stayed pending judicial review, requires states to develop plans to reduce CO2 emissions from existing generating units that use fossil fuels.

- Combined with lower natural gas prices and the extension of renewable tax credits, the CPP accelerates a shift toward less carbon-intensive electricity generation.

- The Reference case includes the CPP and assumes that states select the mass-based limits on CO2 emissions. An alternative case in AEO2017 assumes that the CPP is not implemented.

- AEO2016 included extensive analysis of the CPP and presented several side cases that examined various compliance options available to states.

Reference case energy-related carbon dioxide emissions fall—

U.S. energy-related carbon dioxide emissions (Reference case)
billion metric tons of carbon dioxide

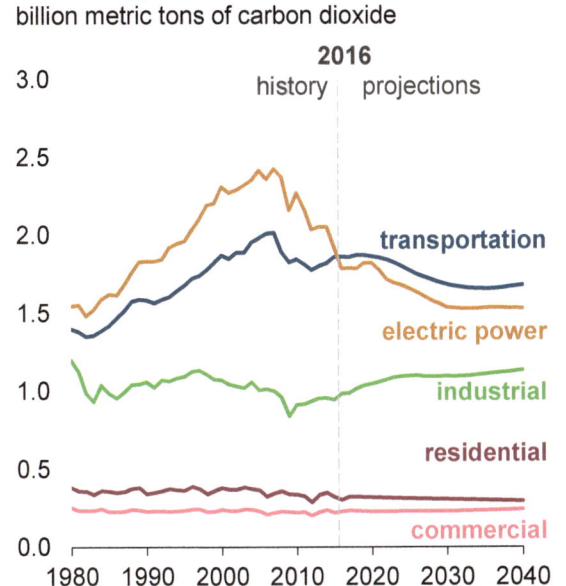

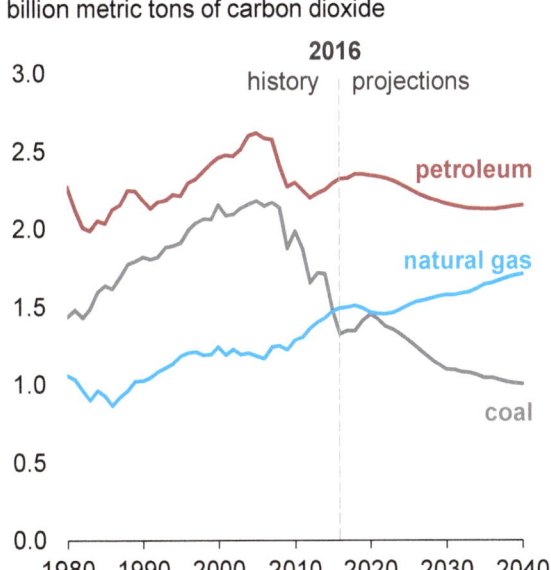

—but at a slower rate than in the recent past

- From 2005 to 2016, energy-related carbon dioxide (CO2) emissions fell at an average annual rate of 1.4%. From 2016 to 2040, energy-related CO2 emissions fall 0.2% annually in the Reference case.

- In the industrial sector, growth in domestic industries, such as bulk chemicals, leads to higher energy consumption and emissions.

- In the electric power sector, coal-fired plants are replaced primarily with new natural gas, solar, and wind capacity, which reduces electricity-related CO2 emissions.

- Direct emissions in the residential and commercial building sectors are largely from space heating, water heating, and cooking equipment. The CO2 emissions associated with the use of electricity in these sectors exceed the direct emissions from these sectors.

- Energy-related CO2 emissions from the transportation sector surpassed those from the electric power sector in 2016. Transportation CO2 emissions remain relatively flat after 2030 as consumption and the carbon intensity of transportation fuels stay relatively constant.

Although population and economic output per capita are assumed to continue rising—

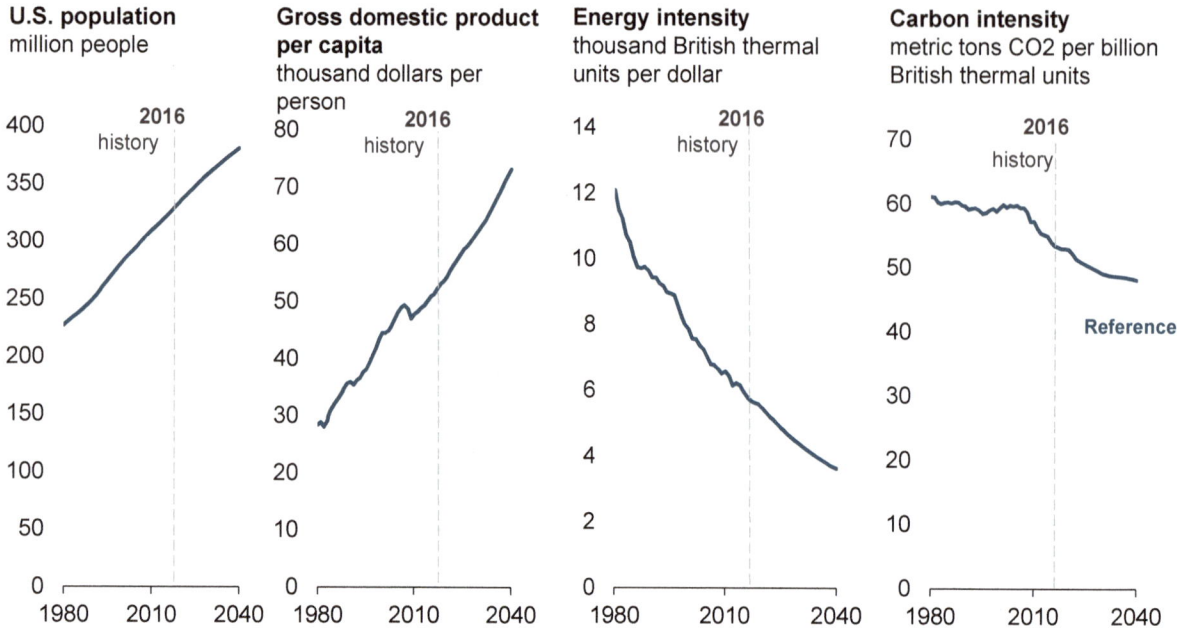

U.S. population
million people

Gross domestic product per capita
thousand dollars per person

Energy intensity
thousand British thermal units per dollar

Carbon intensity
metric tons CO2 per billion British thermal units

—energy intensity and carbon intensity are projected to continue falling in the Reference case

- In the United States, the amount of energy used per unit of economic growth (energy intensity) has declined steadily for many years, while the amount of CO2 emissions associated with energy consumption (carbon intensity) has generally declined since 2008.

- These trends are projected to continue as energy efficiency, fuel economy improvements, and structural changes in the economy all lower energy intensity.

- Carbon intensity declines largely as a result of changes in the U.S. energy mix that reduce the consumption of carbon-intensive fuels and increase the use of low- or no-carbon fuels.

- By 2040, energy intensity and carbon intensity are 37% and 10% lower than their respective 2016 values in the Reference case, which assumes only the laws and regulations currently in place.

Different macroeconomic assumptions address the energy implications of the uncertainty—

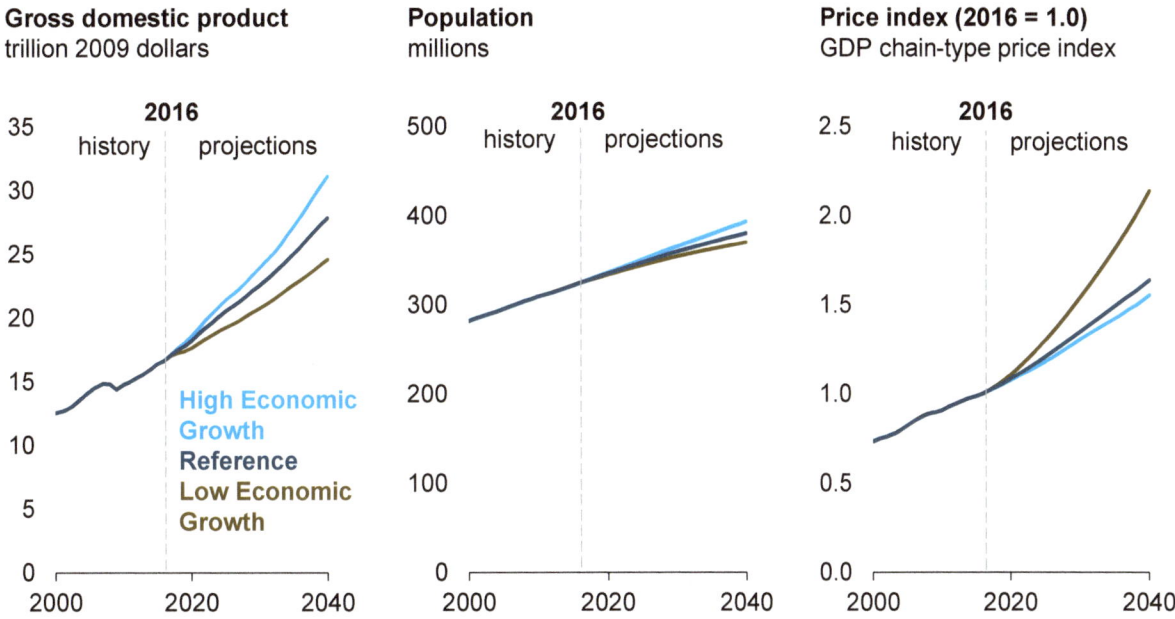

Gross domestic product
trillion 2009 dollars

Population
millions

Price index (2016 = 1.0)
GDP chain-type price index

(Gross domestic product chart legend:)
High Economic Growth
Reference
Low Economic Growth

—surrounding future economic trends

- The Reference, High Economic Growth, and Low Economic Growth cases illustrate three possible paths for U.S. economic growth. The High Economic Growth case assumes higher annual growth and lower annual inflation rates (2.6% and 1.9%, respectively) than in the Reference case (2.2% and 2.1%, respectively), while the Low Economic Growth case assumes lower growth and higher inflation rates (1.6% and 3.2%, respectively).

- In general, higher economic growth (as measured by gross domestic product) leads to greater investment, increased consumption of goods and services, more trade, and greater energy consumption.

- Differences among the cases reflect different expectations for growth in population, labor force, capital stock, and productivity. These changes affect growth rates in household formation, industrial activity, and amounts of travel, as well as investment decisions for energy production.

- All three cases assume smooth economic growth and do not anticipate business cycles or large economic shocks.

Reference case oil prices rise from current levels while natural gas prices remain relatively low—

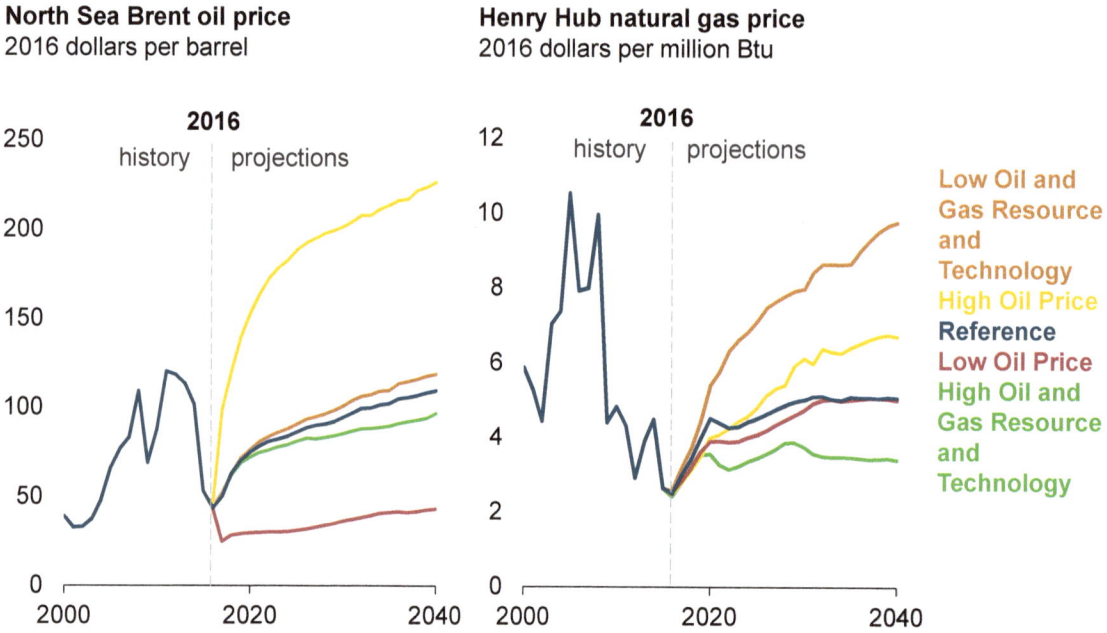

North Sea Brent oil price
2016 dollars per barrel

Henry Hub natural gas price
2016 dollars per million Btu

—price paths in the side cases are very different from those in the Reference case

- In real terms, crude oil prices in 2016 (based on the global benchmark North Sea Brent) were at their lowest levels since 2004, and natural gas prices (based on the domestic benchmark Henry Hub) were the lowest since prior to 1990. Both prices are projected to increase over the projection period.

- Crude oil prices in the Reference case are projected to rise at a faster rate in the near term than in the long term. However, price paths vary significantly across the AEO2017 side cases that differ in assumptions about U.S. resources and technology and global market conditions.

- Natural gas prices in the Reference case also rise and then remain relatively flat at about $5 per million British thermal units (MMBtu) over 2030–40, then rise again over the following decade (not shown on the graph). Projected U.S. natural gas prices are highly sensitive to assumptions about domestic resource and technology explored in the side cases.

United States crude oil and natural gas production depends on oil prices—

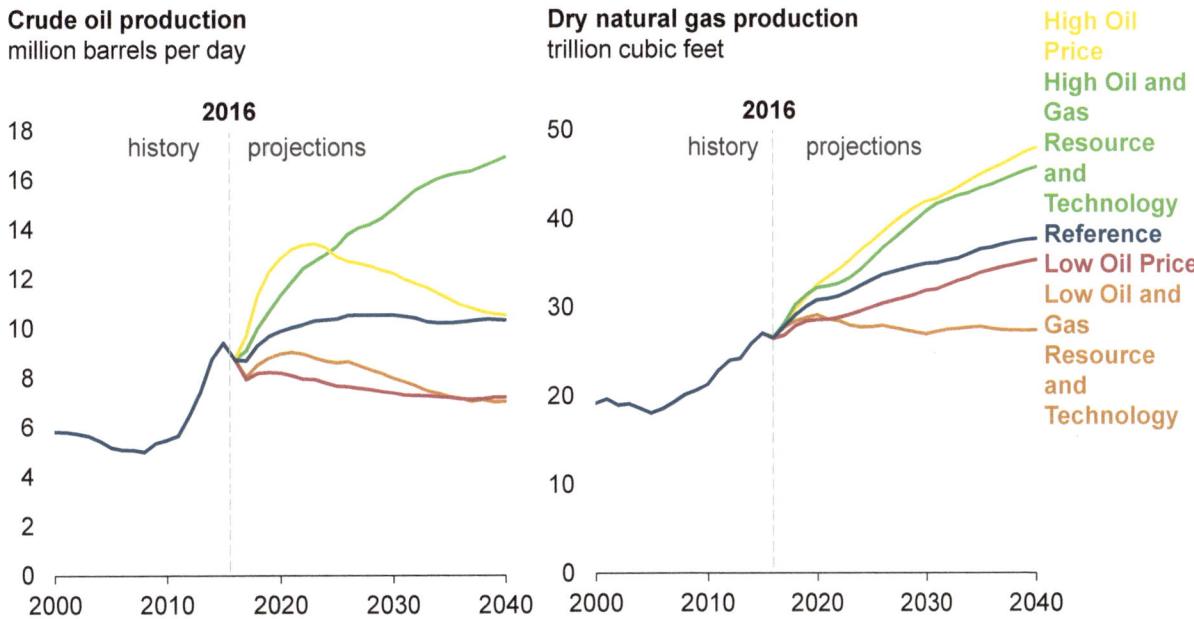

Crude oil production
million barrels per day

2016
history | projections

Dry natural gas production
trillion cubic feet

2016
history | projections

High Oil Price
High Oil and Gas Resource and Technology
Reference
Low Oil Price
Low Oil and Gas Resource and Technology

—as well as resource availability and technological improvements

- Projections of tight oil and shale gas production are uncertain because large portions of the known formations have relatively little or no production history, and extraction technologies and practices continue to evolve rapidly. Continued high rates of drilling technology improvement could increase well productivity and reduce drilling, completion, and production costs.

- In the High Oil and Gas Resource and Technology case, both crude oil and natural gas production continue to grow.

- Crude oil prices affect natural gas production primarily through changes in global natural gas consumption/exports, as well as increases in natural gas production from oil formations (associated gas).

- In the High Oil Price case, the difference between the crude oil and natural gas prices creates more incentive to consume natural gas in energy-intensive industries and for transportation, and to export it overseas as liquefied natural gas, all of which drive U.S. production upward. Without the more favorable resources and technological developments found in the High Oil and Gas Resource and Technology case, U.S. crude oil production begins to decline in the High Oil Price case, and by 2040, production is nearly the same as in the Reference case.

Critical drivers and uncertainty

Various factors influence the model results in AEO2017, including: new and existing laws and regulations, updated data, changing market conditions, and model improvements since AEO2016.

New laws and regulations reflected in the Reference Case

- California state law SB-32, which was passed in 2016, requires statewide greenhouse gas emissions to be 40% below the 1990 level by 2030. This law has cross-cutting effects in California, particularly on electricity and transportation emissions, and also has national implications because of the size of California's energy market.

- The second phase of Federal Greenhouse Gas and Fuel Efficiency standards for medium- and heavy-duty vehicles was issued in 2016. These standards, which ramp up through model year 2027, reduce energy consumption in the transportation sector in the midterm.

Significant data updates

- Data from the 2012 Commercial Buildings Energy Consumption Survey (CBECS) were released in 2016, leading to revised estimates of commercial building mix and energy consumption.

- Updated data on lower battery costs increased EIA's outlook for sales of battery electric vehicles and plug-in hybrid electric vehicles.

Model improvements

- This AEO is the first projection to include model results through 2050, which are available on the AEO page of the EIA website. The graphics in this presentation focus on projections through 2040.

- AEO2017 better captures the dynamics of well productivity that occur when tight oil development moves into less productive areas and as tighter well spacing in established areas diminishes the productivity of each well.

- In contrast to prior AEOs, the AEO2017 Reference case does not assume all nuclear plants that operate through the end of a 60-year period (a 40-year initial operating license plus a 20-year license renewal period) will apply for and receive a subsequent license renewal (SLR) and operate for an additional 20 years. Instead, 25% of reactors reaching age 60 are assumed to retire.

Changing market conditions

- Continuing the trend in previous AEOs, demand for crude oil imports weakens as Lower 48 onshore tight oil development continues to be the main driver of total U.S. crude oil production, accounting for about 60% of cumulative domestic production between 2016 and 2040 in the Reference case.

- Policy-driven economic incentives accelerate renewable generation. With a continued (but reduced) tax credit, solar capacity growth continues throughout the projection period, while tax credits provided for plants entering service until, but no later than 2024, provide incentives for new wind capacity in the near term.

- With solar energy's declining capital costs and solar electricity output that is highest during times of high (on-peak) demand, solar capacity is anticipated to grow throughout the projection period.

EIA will continue to update and refine the market dynamics and technologies in future AEOs, especially with the projection extended to 2050. Ongoing work aims to:

Electric Power

- Energy storage: Improve the representation of energy storage to accommodate multiple grid services including spinning reserve and renewables integration.

- Renewable generation: Include improved representation of intermittent generation resources such as wind and solar. Examine the potential for transmission enhancements to mitigate regional effects of high levels of wind and solar generation. Develop higher resolution time-of-day and seasonal value and operational impact of wind.

- Utility rate structure: Estimate the impact of high levels of distributed photovoltaic generation on utility rate structure.

- Generator retirement: Assess the vintage of the electric generation fleet and potential for future retirements and life extension for all technologies, including existing nuclear, coal, natural gas, and renewable fleets.

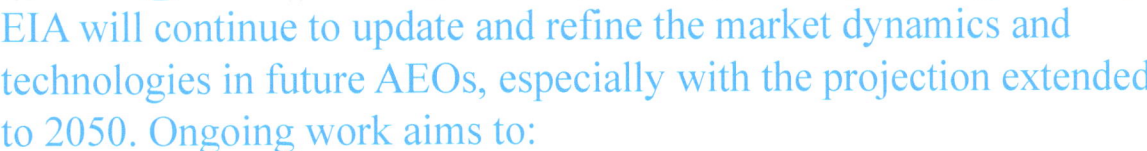

EIA will continue to update and refine the market dynamics and technologies in future AEOs, especially with the projection extended to 2050. Ongoing work aims to:

Liquid Fuels

- Natural gas plant liquids: Re-examine and improve natural gas plant liquids production to allow for changing proportions in produced natural gas over time.

- Technology: Update biofuels and emerging technological assumptions for gas-to-liquids, coal-to-liquids, and carbon sequestration. Improve feedstock curves for all biofuel technologies.

Natural Gas

- Transmission: Improve representation of natural gas market flows with a redesigned NEMS module, allowing for increased flexibility to respond to changing market dynamics (i.e., changing regional flows/bi-directional flow). Improve regional and temporal granularity.

EIA will continue to update and refine the market dynamics and technologies in future AEOs, especially with the projection extended to 2050. Ongoing work aims to:

Transportation

- Technology: Add autonomous vehicle technologies in the transportation sector and consider their implications for on-road fuel economy and total travel demand. Develop the capability to evaluate scenarios where commercial delivery vehicles can operate without human operators and do not require occupant protection features.

- Behavior: Examine the impact of ridesharing programs on travel behavior, including the amount of travel and vehicle choice decisions.

- Fleet mix: Examine determinants of the evolution of the light-duty vehicle fleet mix, which can affect fuel use given the different fuel economy standards for passenger cars and light trucks.

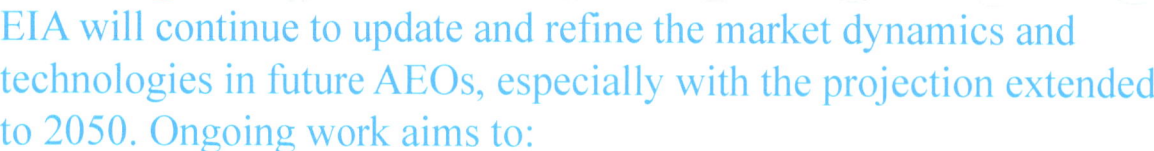

EIA will continue to update and refine the market dynamics and technologies in future AEOs, especially with the projection extended to 2050. Ongoing work aims to:

Buildings

- Distributed generation: Conduct further research and enhance building representation of distributed generation such as photovoltaic, including battery technologies.

- Technology: Review the spread of light emitting diodes and other efficient technologies in buildings. Investigate the adoption of sensor technologies for lights and heating/air conditioning in buildings.

Industrial

- Technology: Incorporate technological change into the industrial model. Apply ongoing technology assessment research in metal-based durables and bulk chemicals to revise energy-intensity projections in those industries.

- Environment: Research the feasibility of carbon capture and storage and implement for carbon-intensive industries such as bulk chemicals, steel, and cement.

Petroleum and other liquids

U.S. crude oil production rebounds from recent lows, driven by continued development of tight oil resources. With consumption flat to down compared to recent history, net crude oil and petroleum product imports as a percentage of U.S. product supplied decline across most cases.

U.S. petroleum product consumption remains below 2005 levels through 2040 in most AEO2017 cases—

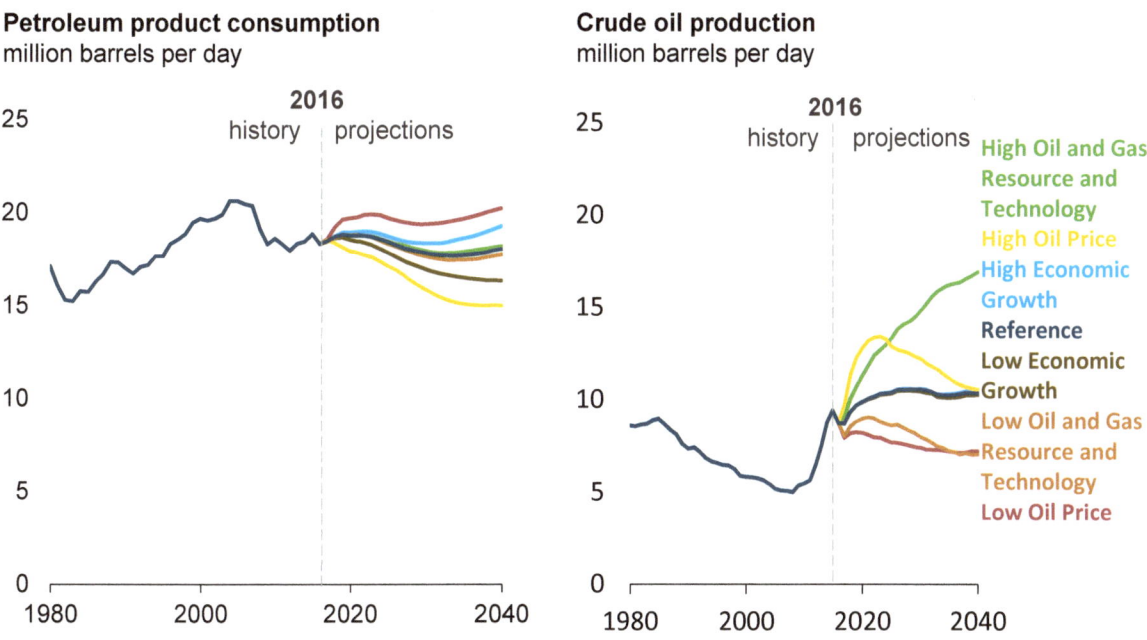

Petroleum product consumption
million barrels per day

Crude oil production
million barrels per day

—while crude oil production rebounds from recent declines

- In all cases, U.S. petroleum consumption is projected to remain below the 2005 level, the highest recorded to date, through 2040.

- Low oil prices result in increased domestic consumption in the Low Oil Price case. Simultaneously, low prices drive down domestic production, resulting in generally higher import levels.

- The domestic wellhead price does not change significantly in the economic growth cases, resulting in consumption that is similar to the Reference case level.

- Reference case U.S. crude oil production is projected to recover from recent declines, as upstream producers increase output because of the combined effects of the rise in prices from recent lows and cost reductions.

- In the Reference case, higher refinery inputs in the near term absorb higher forecast levels of U.S. crude oil production, limiting changes to imports. Eventually, net crude oil imports increase because domestic crude production does not keep pace with refinery inputs as domestic refiners expand product exports.

Tight oil dominates U.S. production in the Reference case—

Crude oil production
million barrels per day

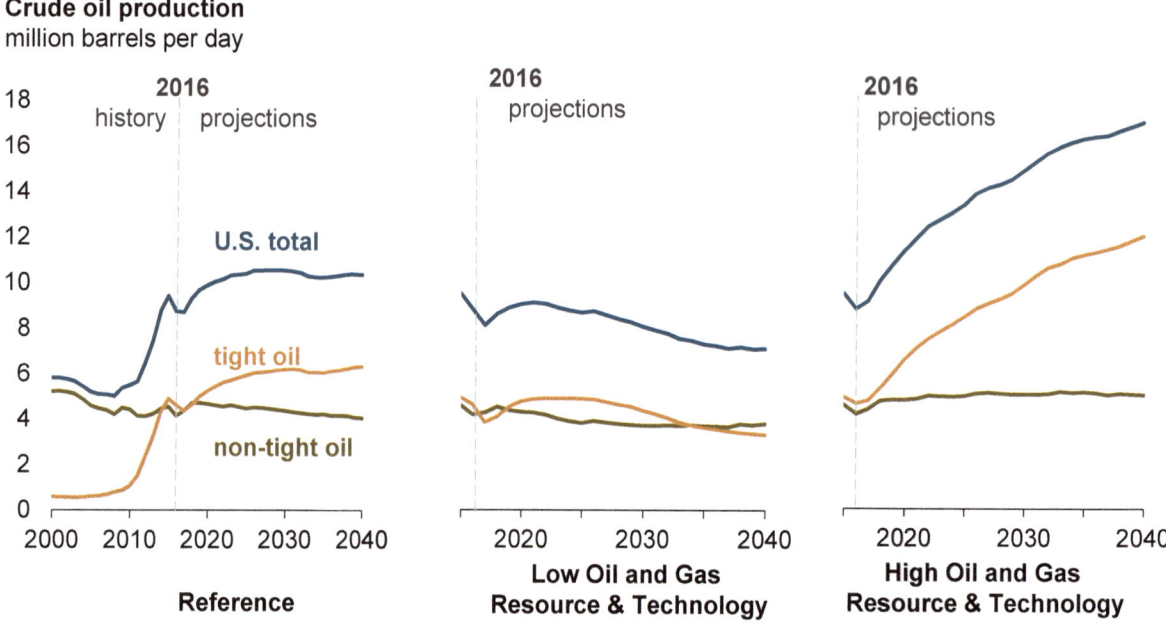

Reference   **Low Oil and Gas Resource & Technology**   **High Oil and Gas Resource & Technology**

—but other types of oil production continue to yield significant volumes

- Despite rising prices, Reference case U.S. crude oil production levels off between 10 and 11 million barrels per day as tight oil development moves into less productive areas and as well productivity gradually decreases.

- Lower 48 onshore tight oil development continues to be the main driver of total U.S. crude oil production, accounting for about 60% of the total cumulative domestic production in the Reference case domestic between 2016 and 2040.

- Announced discoveries in deepwater Gulf of Mexico lead to production increases in the Lower 48 states offshore through 2020. Reference case offshore production then declines until 2034, with the rate of decline slowing through 2040 as production from new discoveries offset declines in legacy fields.

- In the High Oil and Gas Resource and Technology case, higher well productivity reduces development and production costs per unit, resulting in more resource development than in the Reference case. These assumptions are based on higher initial estimated ultimate recovery per well, larger volumes of onshore Lower 48 tight oil and shale gas resources, and higher rates of long-term technology improvement.

The Southwest and Dakotas/Rocky Mountains regions lead growth in tight oil production in the Reference case—

Lower 48 onshore crude oil production by region (Reference case)
million barrels per day

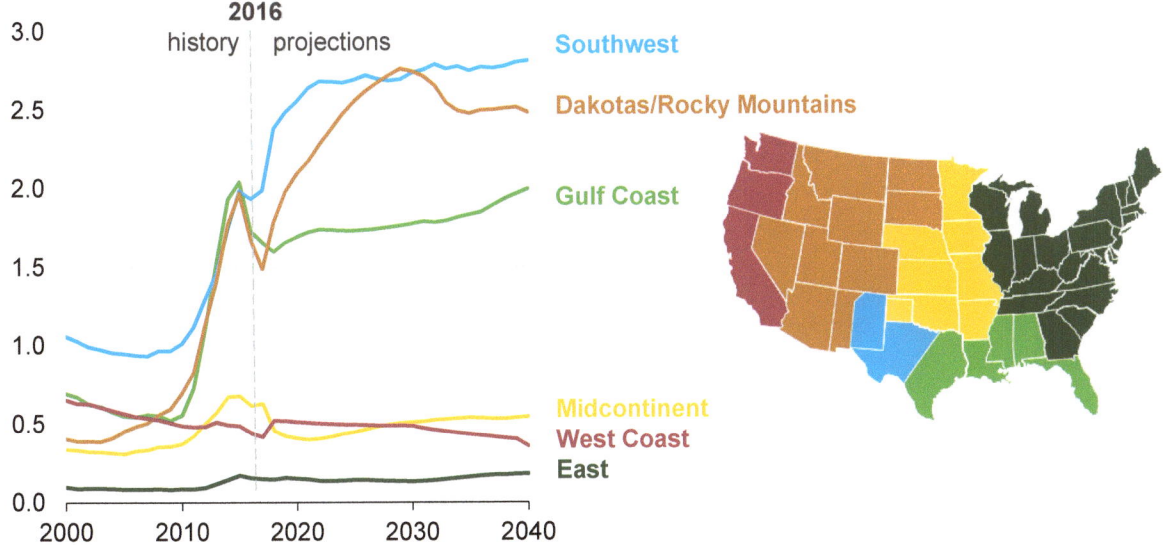

—and the Gulf Coast region remains an important contributor to overall production levels

- Growth in Lower 48 onshore crude oil production is projected to occur mainly in the Southwest, Dakotas/Rocky Mountains, and Gulf Coast regions.

- Growth in crude oil production in the Southwest is supported by increases in the Permian basin, which includes both tight and non-tight formations.

- Growth in the Dakotas/Rocky Mountains crude oil production is driven by increased production from the Bakken play, which is exclusively tight oil.

- Production in the Gulf Coast region, primarily from the Eagle Ford and Austin Chalk plays, increases throughout most of the projection period.

In most cases, the United States remains a net petroleum importer—

Petroleum net imports as a percentage of products supplied
percent

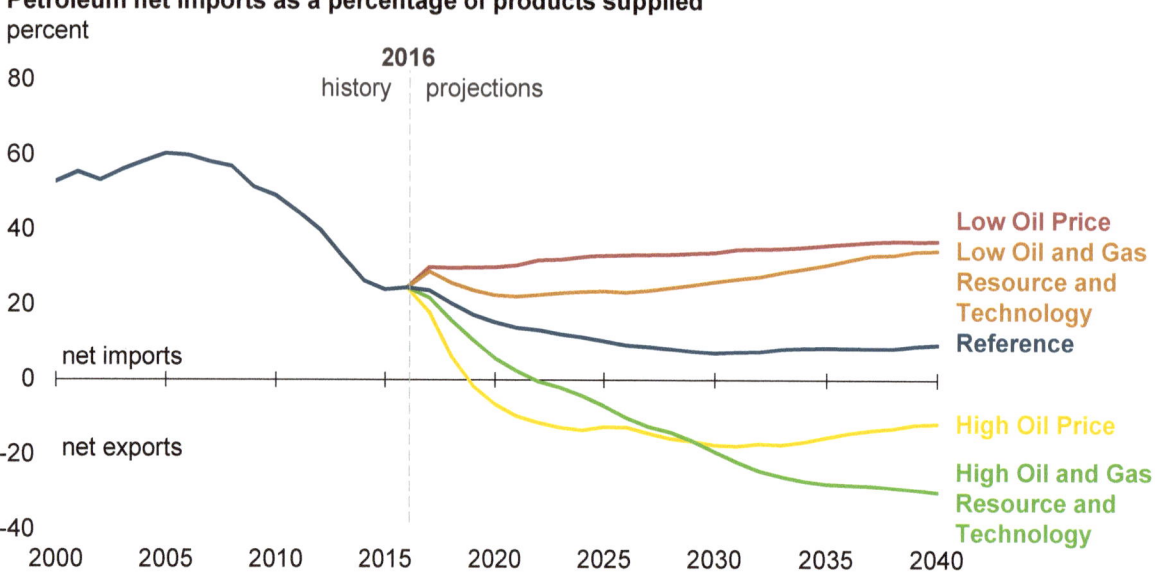

—but in the High Oil Price and the High Oil and Gas Resource and Technology cases, the United States becomes a net exporter

- In the Reference case, net crude oil and petroleum product imports as a percentage of U.S. product supplied fall through 2030.

- The Low Oil Price case results in lower U.S. crude oil production because of the lack of economic incentive for producers to drill in higher-cost tight oil formations and offshore crude oil reserves. Relatively lower prices in this case result in higher domestic product demand that promotes higher crude oil and petroleum product imports.

- In the High Oil Price case, high crude oil prices lead to increased U.S. crude oil production from higher-cost production areas and result in lower domestic petroleum product demand, which leads to lower product imports.

- In the High Oil and Gas Resource and Technology case, U.S. crude oil and petroleum liquids exports are higher compared with the Reference case.

U.S. motor gasoline consumption and exports are sensitive to changes in prices—

Motor gasoline retail prices

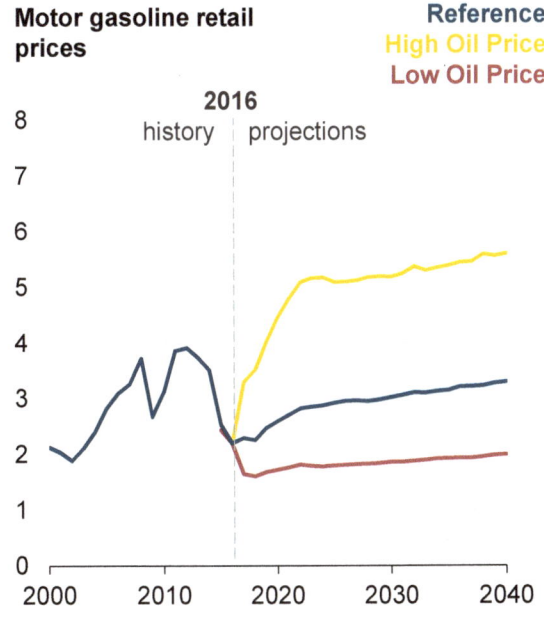

Reference
High Oil Price
Low Oil Price

Motor gasoline consumption and gross exports
million barrels per day

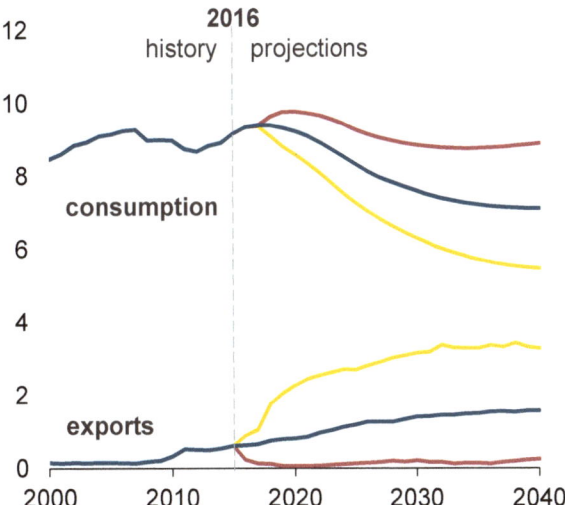

—although efficiency improvements result in declining consumption across all cases

- U.S. average retail prices for motor gasoline are driven largely by changes in crude oil prices because crude oil is the main input used to produce motor gasoline.

- Improvements in vehicle fuel efficiency contribute to falling U.S. motor gasoline consumption, while high levels of refinery output result in continued growth of motor gasoline exports through 2040.

- In the Low Oil Price case, greater domestic motor gasoline consumption and lower domestic crude oil production results in lower exports of motor gasoline.

- The High Oil Price case results in lower domestic motor gasoline consumption and greater exports, reflecting the domestic gasoline demand response to higher prices as well as the U.S. refining industry's competitive advantage.

Natural gas

Across most cases, natural gas production increases despite relatively low and stable natural gas prices, supporting higher levels of domestic consumption and natural gas exports. Projections are sensitive to resource and technology assumptions.

U.S. natural gas consumption increases across most cases through most of the projection period—

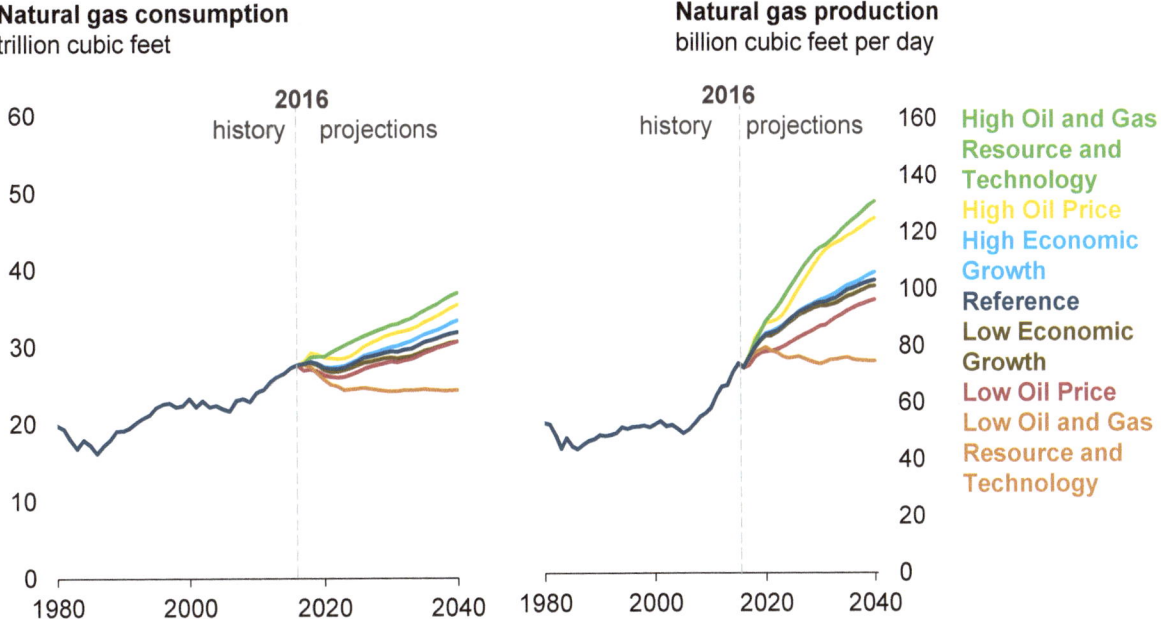

Natural gas consumption
trillion cubic feet

Natural gas production
billion cubic feet per day

High Oil and Gas Resource and Technology
High Oil Price
High Economic Growth
Reference
Low Economic Growth
Low Oil Price
Low Oil and Gas Resource and Technology

—and in combination with growing net exports, supports production growth

- In the Reference case, natural gas production over the 2016–20 period is projected to grow at about the same rapid rate (nearly 4% annual average) as it has since 2005. Since 2005, technologies to more efficiently produce natural gas from shale and tight formations have driven prices down, spurring growth in consumption and net exports.

- Beyond 2020, natural gas production in the Reference case is projected to grow at a lower rate (1.0% annual average) as net export growth moderates, domestic natural gas use becomes more efficient, and prices slowly rise. Rising prices are moderated by assumed advances in oil and natural gas extraction technologies.

- Near-term production growth is supported by large, capital-intensive projects, such as new liquefaction export terminals and petrochemical plants, built in response to low natural gas prices.

- Despite decreasing in the near term, in all cases, other than the Low Oil and Gas Resource and Technology case, U.S. natural gas consumption is expected to increase during much of the projection period.

Natural gas prices are projected to increase—

Natural gas spot price at Henry Hub
2016 dollars per million British thermal units

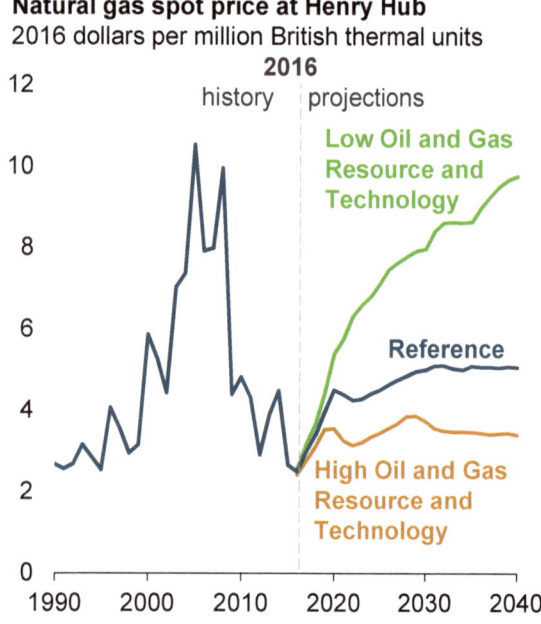

Dry natural gas production
trillion cubic feet

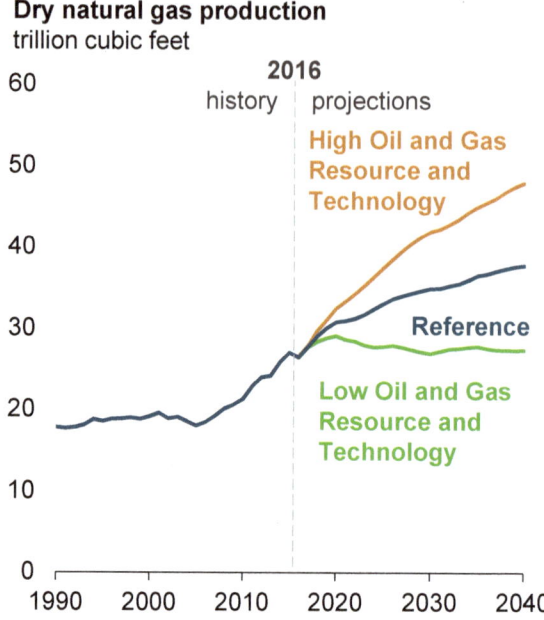

—and are sensitive to the availability of new technology and resources

- The range of projected Henry Hub natural gas prices depends on the assumptions about the availability of oil and natural gas resources and drilling technology.

- In the Reference case, the natural gas spot prices at the U.S. benchmark Henry Hub in Louisiana rise because of increased drilling levels, production expansion into less prolific and more expensive-to-produce areas, and demand from both petrochemical and liquefied natural gas export facilities.

- Reference case prices rise modestly from 2020 through 2030 as electric power consumption increases; however, natural gas prices stay relatively flat after 2030 as technology improvements keep pace with rising demand.

- In the High Oil and Gas Resource and Technology case, lower costs and higher resource availability allow for increased levels of production at lower prices, increasing domestic consumption and exports.

- In the Low Oil and Gas Resource and Technology case, prices near historical highs drive down domestic consumption and exports.

U.S. natural gas production growth is the result of continued development of shale gas and tight oil plays—

Dry natural gas production by type
trillion cubic feet

billion cubic feet per day

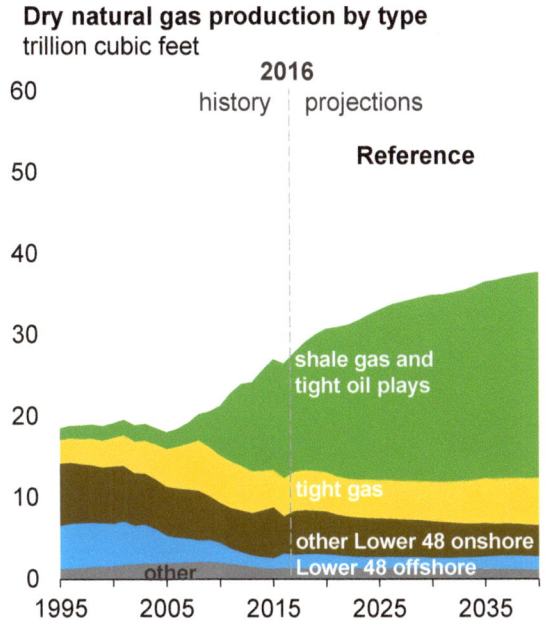

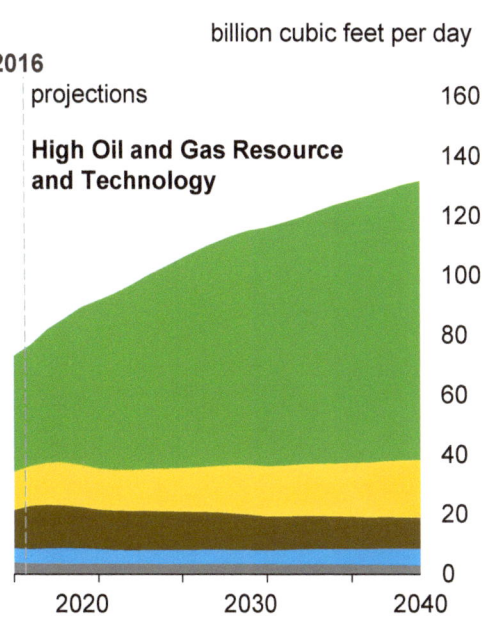

—which account for nearly two-thirds of natural gas production by 2040

- Production from shale gas and associated gas from tight oil plays is the largest contributor to natural gas production growth, accounting for nearly two-thirds of total U.S. production by 2040 in the Reference case.

- Tight gas production is the second-largest source of domestic natural gas supply in the Reference case, but its share falls through the late-2020s as the result of growing development of shale gas and tight oil plays.

- As new discoveries offset declines in legacy fields, offshore natural gas production in the United States increases over the projection period.

- Production of coalbed methane generally continues to decline through 2040 because of unfavorable economic conditions for producing that resource.

Plays in the East lead production of U.S. natural gas from shale resources in the Reference case—

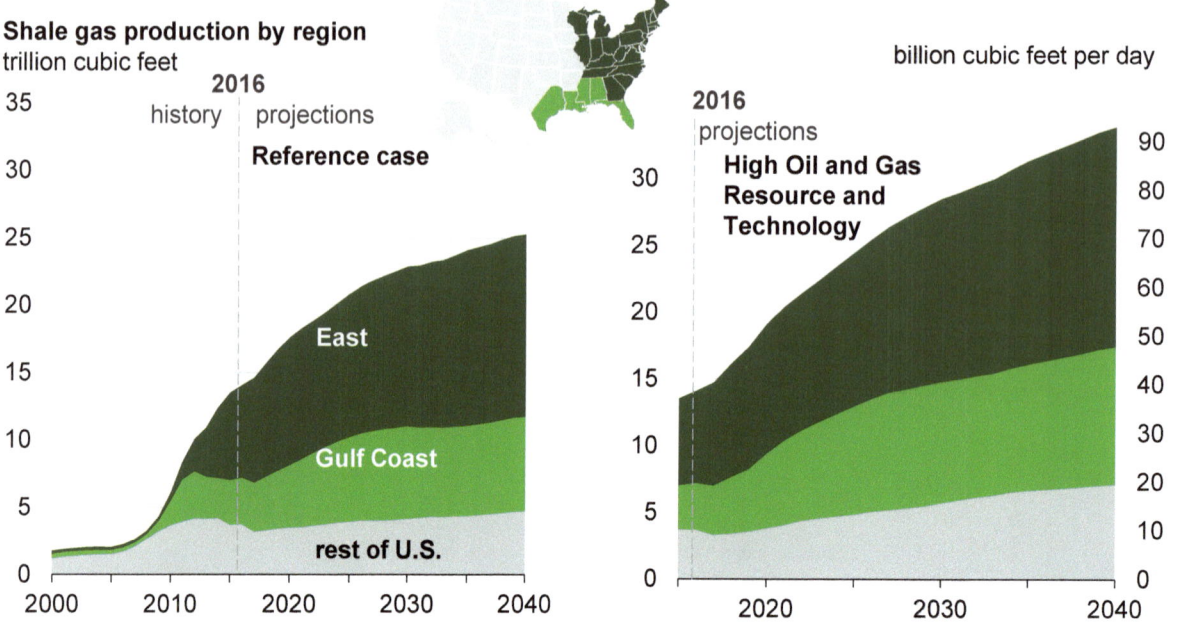

Shale gas production by region
trillion cubic feet

billion cubic feet per day

—but Gulf Coast onshore production also grows

- Continued development of the Marcellus and Utica plays in the East is the main driver of growth in total U.S. shale gas production and the main source of total U.S. dry natural gas production.

- Production from the Eagle Ford and Haynesville plays along the Gulf Coast is a secondary contributor to domestic dry natural gas production, with production largely leveling off in the 2030s.

- Continued technological advancement and improvement in industry practices is expected to lower costs and to increase the expected ultimate recovery per well. These changes have a significant cumulative effect in plays that extend over wide areas and have large undeveloped resources (Marcellus, Utica, and Haynesville).

Increasing demand from industrial and electric power markets drive rising domestic consumption of natural gas in the Reference case—

Natural gas consumption by sector
trillion cubic feet

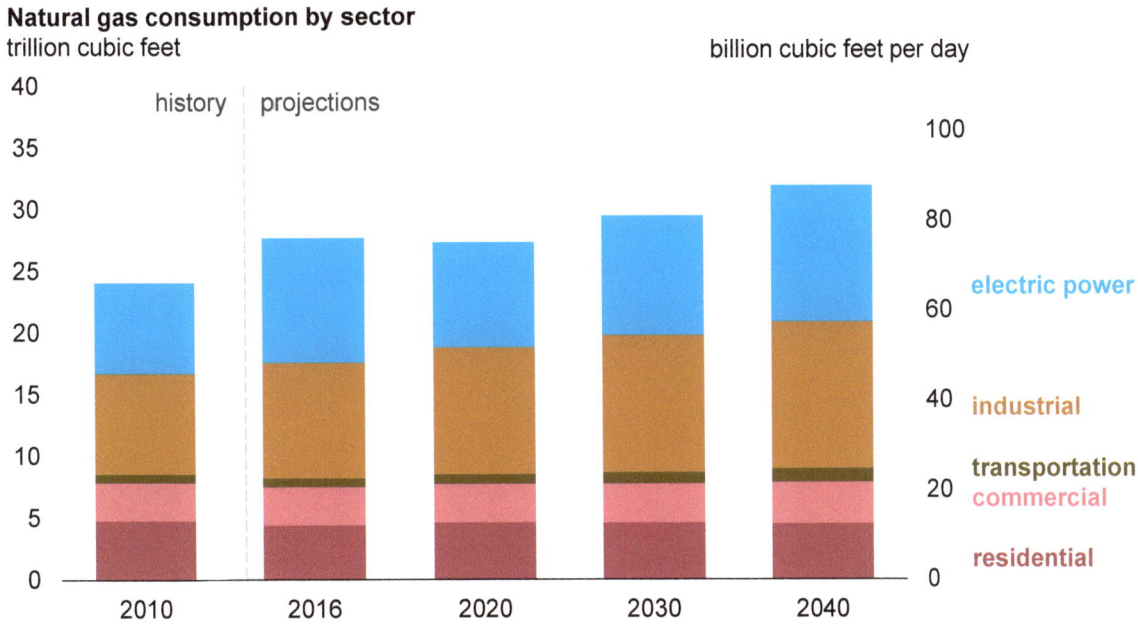

—with comparatively little growth in the residential and commercial sectors

- The industrial sector is the largest consumer of natural gas during most years in the Reference case projections. Major natural gas consumers include the petrochemical industry (where natural gas is used as a feedstock in the production of methanol, ammonia, and fertilizer), other energy-intensive industries that use natural gas for heat and power, and liquefied natural gas producers.

- After a brief near-term decline attributable to strong growth in renewables generation and price competition with coal, natural gas used for electric power generation generally increases after 2020. In particular, the Clean Power Plan (CPP) and the scheduled expiration of renewable tax credits in the mid-2020s result in an increase in the electric power sector's natural gas use. Natural gas consumption in the electric power sector is about 6% higher in the Reference case in 2040 than the No CPP case.

- Natural gas consumption in the residential and commercial sectors remains largely flat as a result of efficiency gains that balance increases in the number of housing units and commercial floor space.

- Although natural gas use rises in the transportation sector, it remains a small share of both total natural gas consumption and transportation fuel demand.

U.S. LNG export levels vary across cases and reflect both the level of global demand—

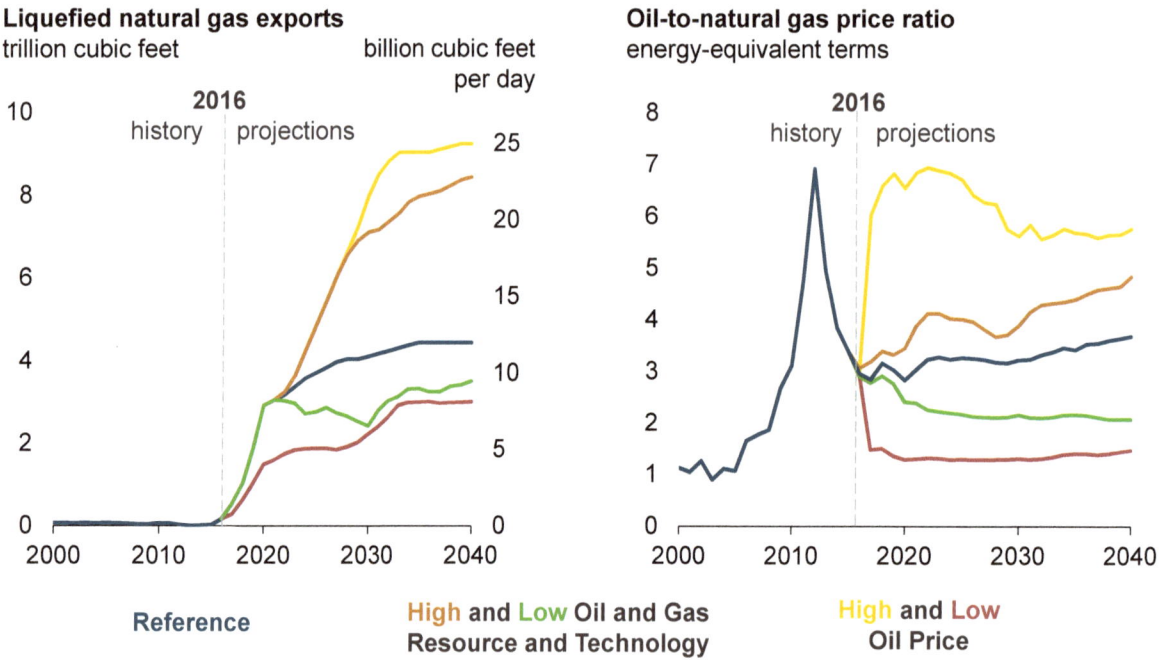

Liquefied natural gas exports
trillion cubic feet

billion cubic feet
per day

2016
history | projections

Reference

High and Low Oil and Gas
Resource and Technology

Oil-to-natural gas price ratio
energy-equivalent terms

2016
history | projections

High and Low
Oil Price

—and the difference between domestic and global natural gas prices, with the latter more heavily influenced by oil prices

- Currently, most liquefied natural gas (LNG) is traded under oil price-linked contracts, in part because oil can substitute for natural gas in industry and for power generation. However, as the LNG market expands, contracts are expected to change, weakening their ties to oil prices.

- When the oil-to-natural gas price ratio is highest, as in the High Oil Price case, U.S. LNG exports are at their highest levels. Demand for LNG generally increases as consumers move away from petroleum products, and LNG produced in the United States has the advantage of domestic spot prices that are less sensitive to global oil prices than supplies from other sources. In the Low Oil Price case, LNG exports from the United States are at their lowest levels throughout the projection period.

- In the High Oil and Gas Resource and Technology case, low U.S. natural gas prices make U.S. LNG exports competitive relative to other suppliers. Conversely, higher U.S. natural gas prices in the Low Oil and Gas Resource and Technology case result in lower U.S. LNG exports.

Increased natural gas trade is dominated by liquefied natural gas exports in the Reference case—

Natural gas trade
trillion cubic feet

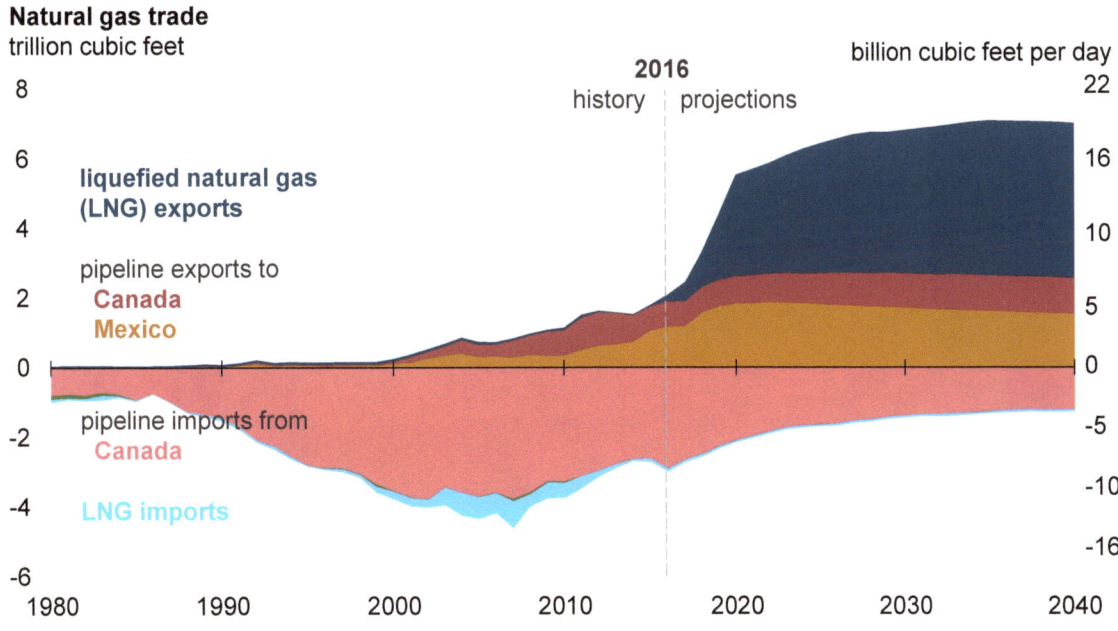

—while pipeline imports into the United States continue to decline

- In the Reference case, liquefied natural gas (LNG) is projected to dominate U.S. natural gas exports by the early-2020s. The first LNG export facility in the Lower 48, Sabine Pass, began operations in 2016, and four more LNG export facilities are scheduled to be completed by 2020.

- After 2020, U.S. exports of LNG grow at a more modest rate as U.S.-sourced LNG becomes less competitive in global energy markets.

- U.S. natural gas exports to Mexico continue to rise in the short term as pipeline infrastructure currently under development allows for rising exports to meet Mexico's increased demand for natural gas to fuel electric power generation.

- U.S. imports of natural gas from Canada, primarily from the West where most of Canada's natural gas is produced, continue to decline, while U.S. exports to Canada—primarily to the East—continue to increase because of Eastern Canada's proximity to abundant natural gas resources in the Marcellus basin.

Electricity

As demand grows modestly, the primary driver for new capacity in the Reference case is the retirement of older, less efficient fossil fuel units—largely spurred by the Clean Power Plan (CPP)—and the near-term availability of renewable energy tax credits. Even if the CPP is not implemented, low natural gas prices and the tax credits result in natural gas and renewables as the primary sources of new generation capacity. The future generation mix is sensitive to the price of natural gas and the growth in electricity demand.

Fuel prices and current laws and regulations drive growing shares of renewables and natural gas in the electricity generation mix—

U.S. net electricity generation from select fuels
billion kilowatthours

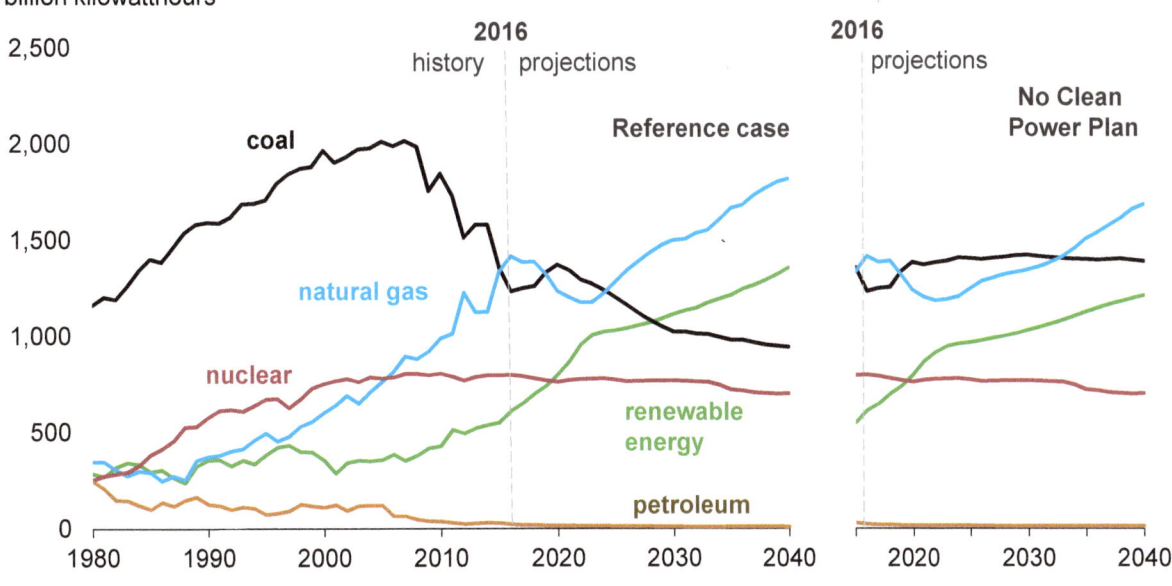

—as coal's share declines over time in the Reference case

- Fuel prices drive near-term natural gas and coal shares. As natural gas prices rebound from their 20-year lows which occurred in 2016, coal regains a larger generation share over natural gas through 2020.

- Federal tax credits drive near-term growth in renewable generation, displacing growth in natural gas.

- In the longer term, policy (Clean Power Plan, renewables tax credits, and California's SB32) and unfavorable economic conditions compared with natural gas and renewables result in declining coal generation and growing natural gas and renewables generation in the Reference case.

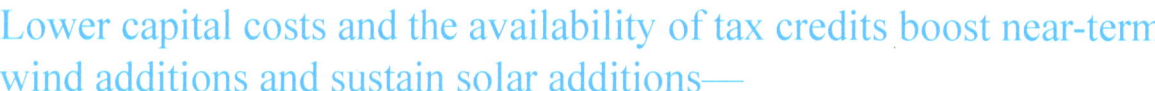

Lower capital costs and the availability of tax credits boost near-term wind additions and sustain solar additions—

Annual electricity generating capacity additions and retirements (Reference case)
gigawatts

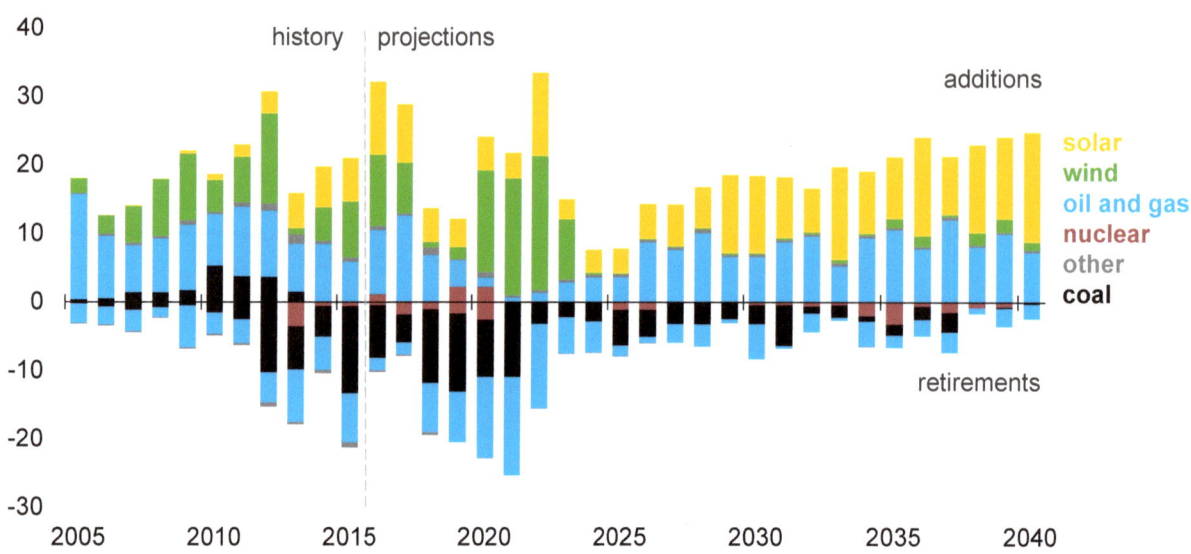

—whereas coal-fired unit retirements in the Reference case are driven by low natural gas prices and the Clean Power Plan

- In the Reference case, nearly 70 gigawatts (GW) of new wind and solar photovoltaic (PV) capacity is added over 2017–21, encouraged by declining capital costs and the availability of tax credits.

- Most of the wind capacity used to comply with the Clean Power Plan (CPP) is built prior to the scheduled expiration of the production tax credit for wind plants coming online by the end of 2023, although wind is still likely to be competitive without the tax credits.

- Continued retirements of older, less efficient fossil fuel units under the CPP support a consistent market for new generating capacity throughout the projection period.

- After 2030, new generation capacity additions are split primarily between solar and natural gas, with solar capacity representing more than 50% of new capacity additions in the Reference case between 2030 and 2040.

Natural gas resource availability affects prices that plays a critical role in determining the mix of coal, natural gas, and renewable generation—

Electricity generation from selected fuels
billion kilowatthours

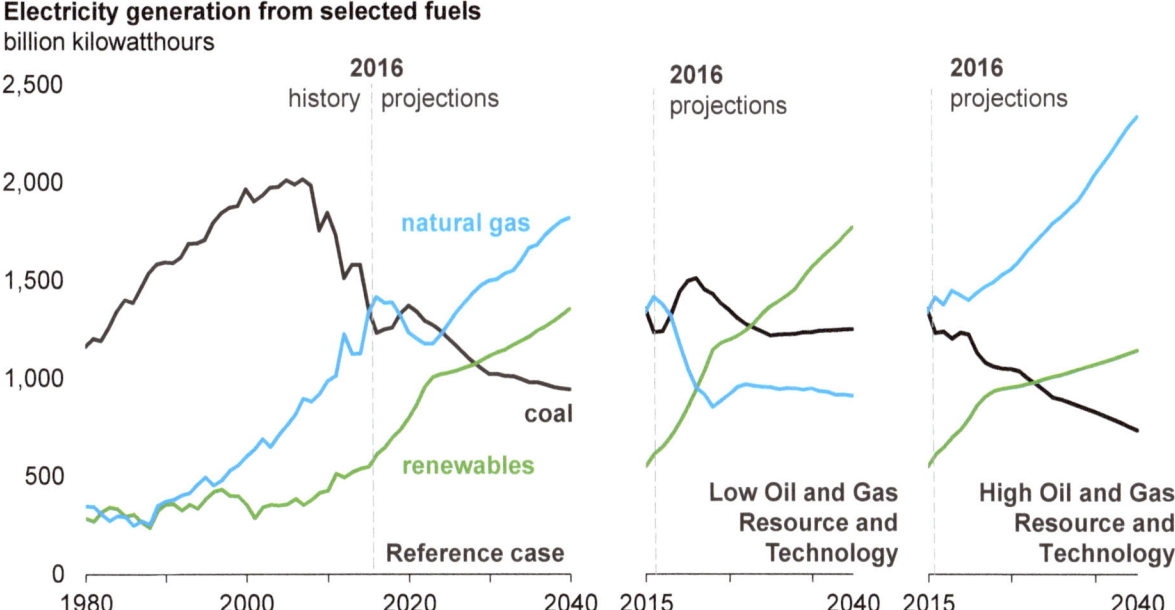

—as seen in the resource and technology cases

- Lower natural gas prices, which occur in the High Oil and Gas Resource and Technology case, lead to natural gas-fired electricity generation displacing coal-fired generation. In this case, and relative to the Reference case, natural gas maintains its market-share lead over coal through 2040, and it displaces some renewables market share relative to the Reference case.

- Higher natural gas prices, which occur in the Low Oil and Gas Resource and Technology case, favor growth of renewables. Relative to the Reference case, coal-fired generation regains market share from natural gas in the near term, but because of carbon emission limits imposed by the Clean Power Plan, renewables ultimately gain a larger market share.

Electricity use continues to increase—

Electricity use by end-use demand sector
billion kilowatthours

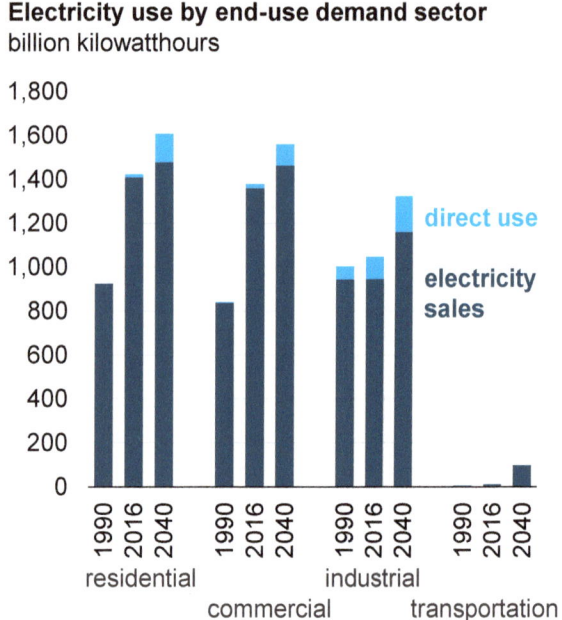

Electricity use growth rate
percent growth (three-year rolling average)

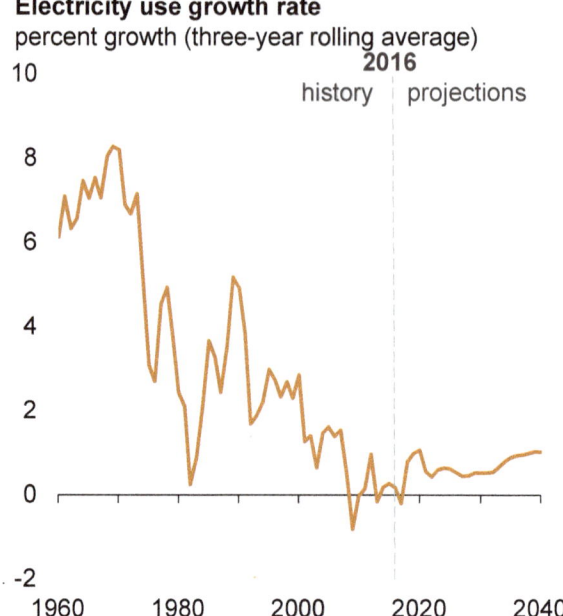

—but the rate of growth remains lower than historic averages in the Reference case

- In recent history, the growth in electricity demand has slowed as older equipment was replaced with newer, more efficient stock, as efficiency standards were implemented and technology change occurred, particularly in lighting and other appliances. The demographic and economic factors driving this trend included slowing population growth and a shifting economy toward less energy-intensive industries.

- While growth in the economy and electricity demand remain linked, historically the linkage has continued to shift toward much slower electricity demand growth relative to economic growth.

- Growth in electricity demand, while relatively low historically, begins to rise slowly across the projection period as demand for electric services is only partially offset by regulatory compliance and efficiency gains in electricity-using equipment.

- Growth in direct use generation above growth in sales is primarily the result of the adoption of rooftop photovoltaic (PV) and natural gas-fired combined heat and power (CHP).

Wind and solar generation become the predominant sources of renewable generation in the Reference case—

Renewable electricity generation (Reference case)
billion kilowatthours

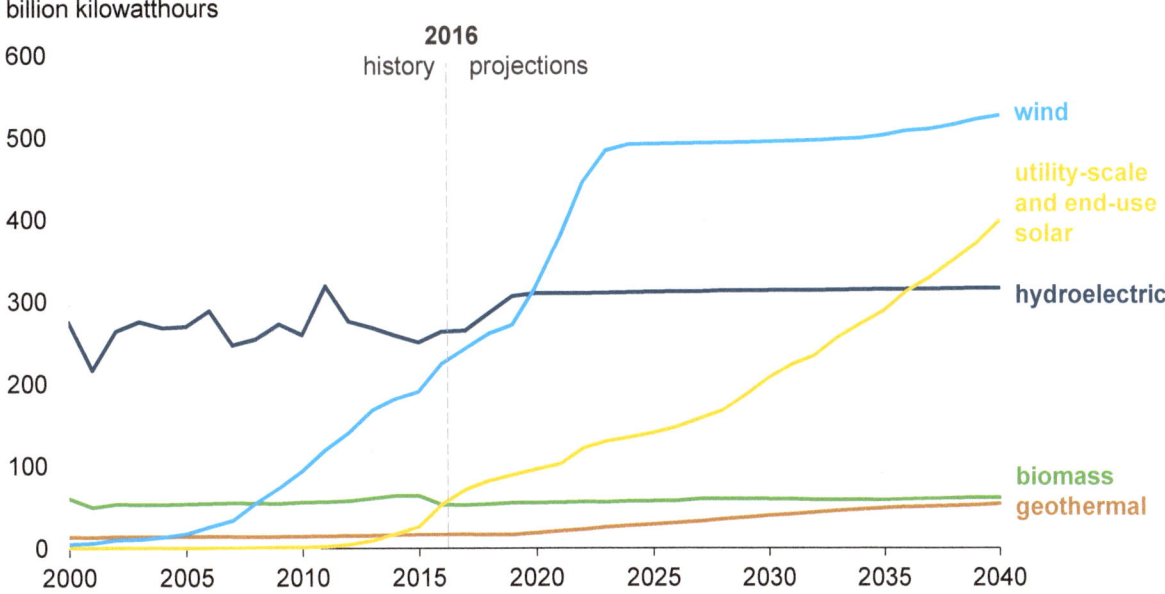

—with each surpassing hydroelectric generation

- The Clean Power Plan (CPP) and state-defined Renewable Portfolio Standards (RPS) increase demand for wind and solar electricity generation throughout the projection period.

- The scheduled expiration of production tax credits encourages an increase in wind capacity additions ahead of CPP implementation. While many wind projects would be economic without the tax credits, most of the profitable wind capacity will be added to take advantage of the tax credits prior to their expiration.

- Substantial cost reductions, performance improvements, and a permanent 10% investment tax credit support solar generation growth throughout the projection period.

- Some geothermal resources are also competitive sources of new generation, but these lowest-cost resources are geographically limited and are only expected to be exploited slowly.

Most electric generation from solar resources comes from utility-scale installations—

Solar electricity generation (Reference case)
billion kilowatthours

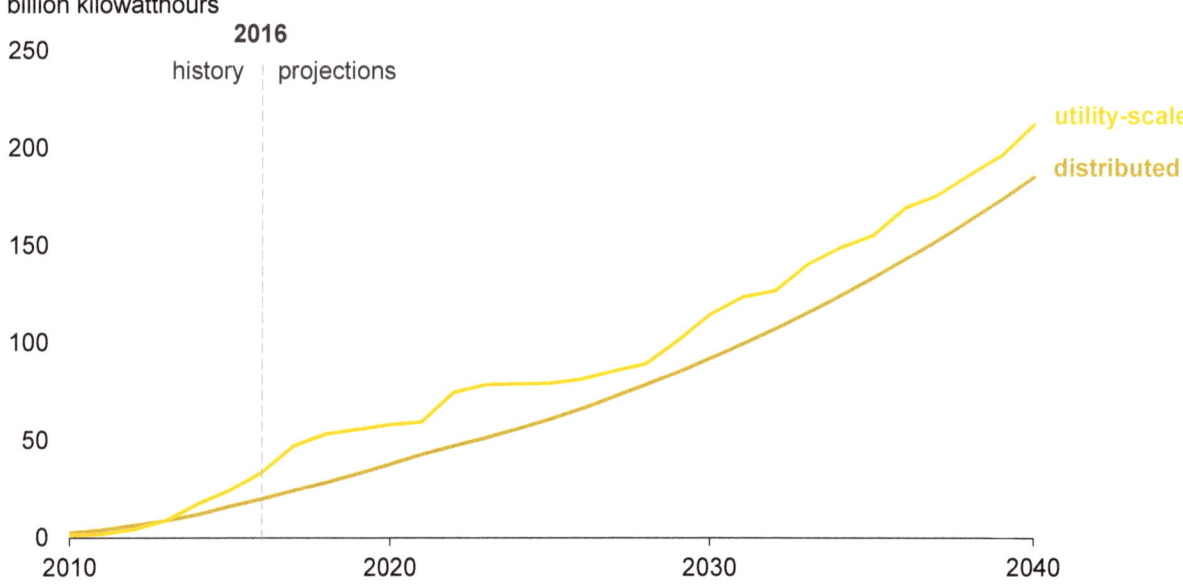

—but generation from distributed photovoltaics is a significant contributor

- Although utility-scale photovoltaic (PV) generation typically costs less than distributed PV, in some circumstances distributed PV remains economically attractive. Distributed PV competes against higher retail electricity prices, which do not necessarily reflect time-of-day or seasonal variation in the cost of electricity.

- With a continued (but reduced) tax credit, declining costs, and on-peak generation profile, both utility and distributed solar builds occur throughout the projection period.

- AEO2017 projections include higher time-of-day and seasonal resolution of both utility-scale and distributed solar output as compared to AEO2016, as well as higher geographic resolution (at the ZIP code level) of distributed solar. The net result of these model changes is to reduce projected utility-scale solar generation and increase distributed solar generation, although not to the same degree.

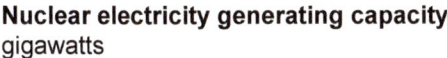

Assumptions about license renewals in AEO2017 increase nuclear retirements—

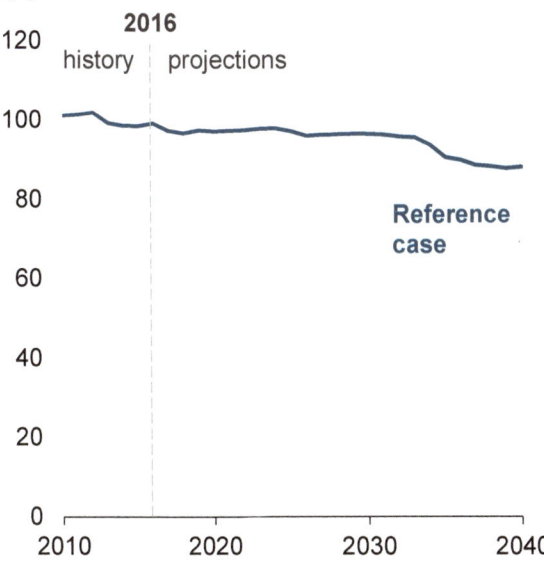

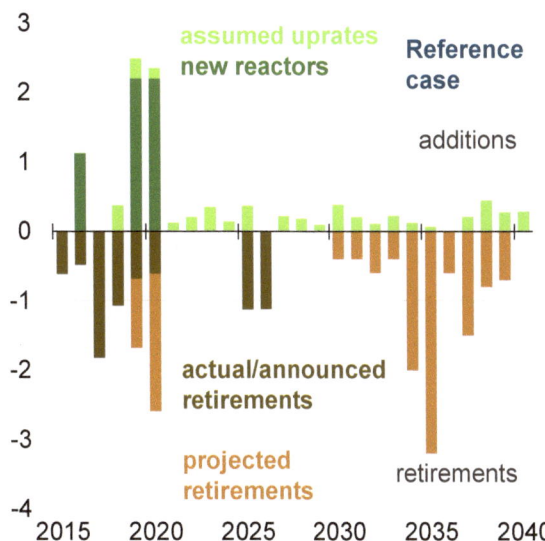

—leading to net nuclear capacity decreases

- No new, unannounced nuclear capacity is added in the Reference case over the projection period because of the combination of low natural gas prices, higher renewables penetration, low electricity load growth, and relatively high capital costs.

- New capacity additions are limited to reactors under construction from 2017 onward and to projected uprates at existing reactors. From 2018 through 2040, 4.7 gigawatts (GW) of additional capacity at existing units is projected to come online, based on an assessment of the remaining uprate potential.

- A significant reduction in nuclear capacity occurs because of 6.4 GW of total announced retirements; 3.0 GW of projected retirements in 2019–20 to address near-term, market uncertainty; and approximately 10.6 GW of long-term retirements through 2040 to address the uncertainty of reactors achieving a subsequent license renewal. As many nuclear plants reach the 60-year subsequent license renewal decision after 2040, retirements continue, with another 11.7 GW of nuclear capacity projected to retire by 2050.

- All nuclear plant retirements other than those already announced were modeled as capacity reductions for the regional nuclear fleets (i.e., as generic derates), rather than as retirements of specific plants.

Coal production decreases—

Coal production
million short tons

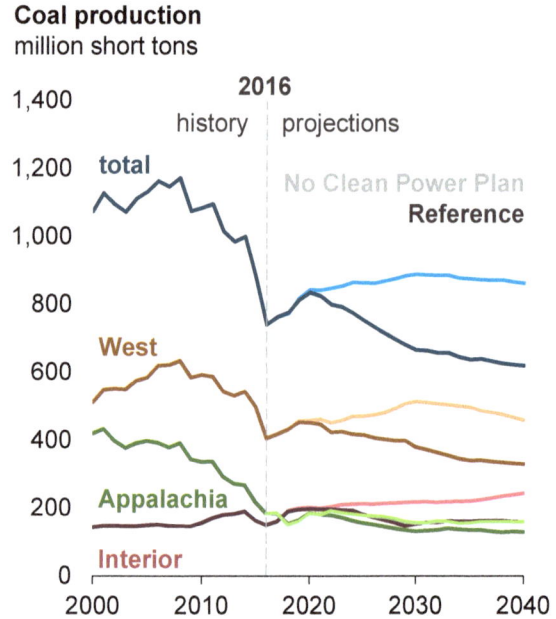

Coal consumption in electric power sector
million short tons

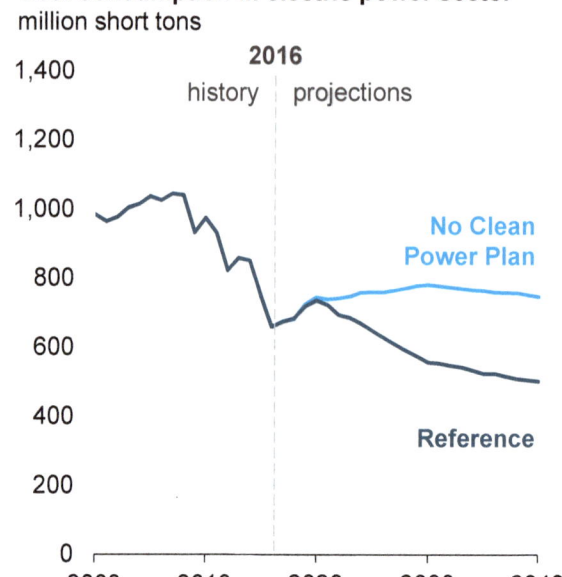

—primarily in the Western region

- The impacts of the Clean Power Plan (CPP) are not shared equally across the major coal supply regions because of differences in coal quality, regional natural gas and coal prices, and how the electricity markets served by each region are affected with respect to coal retirements and renewables penetration.

- Coal production increases through 2020 to more than 800 million short tons in the Reference case as a projected rise in natural gas prices improves the competitiveness of existing coal generating units.

- After 2020, coal production in the Reference case declines, reaching nearly 620 million short tons per year in 2040, which is lower than the over 850 million short tons per year projected to be produced in 2040 in the No CPP case.

- The Interior region market share grows from 20% of U.S. coal production in 2016 to 26% by 2040, with Appalachia and Western production losing market share in both the Reference and No CPP cases.

- Coal production declines gradually after 2030 in the Reference case as retiring nuclear capacity is replaced, in part, by natural gas-fired electricity generation, requiring a reduction in existing carbon-emitting generation to maintain the CPP emission cap.

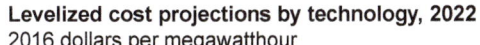

Including available federal tax credits, wind and solar units will be among the most competitive sources of new generation in 2022—

Levelized cost projections by technology, 2022
2016 dollars per megawatthour

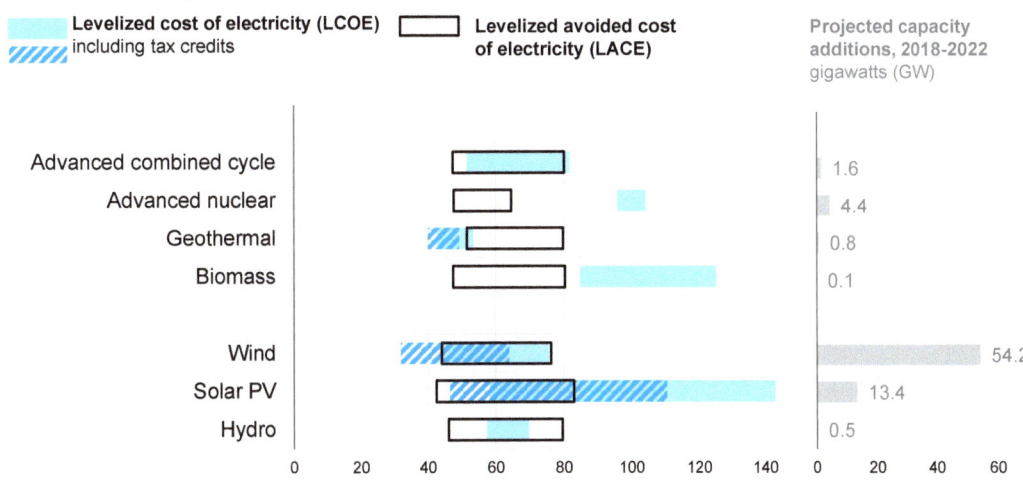

Source: *U.S. Energy Information Administration, Levelized Cost and Levelized Avoided Cost of New Generation Resources in the Annual Energy Outlook 2017*
Note: *Capacity additions include planned and unplanned additions.*

—when levelized costs of electricity and levelized avoided costs of electricity are considered

- Comparisons of levelized cost of electricity (LCOE) across technologies can be misleading as different technologies serve different market segments.

- Levelized avoided cost of electricity (LACE) can be used to compare the cost (LCOE) of an electricity generation resource against the value (LACE) of the electricity generation and capacity that it displaces.

- Wind plants entering service in 2022 that started construction in 2018 will receive an inflation-adjusted $14/MWh federal production tax credit; solar plants entering service in 2022 will receive a 26% investment tax credit, assuming a two-year construction lead time.

- See more information in EIA's LACE/LCOE report on EIA's website.

The value of energy (LACE) for wind and solar is more sensitive to differences in policy and market assumptions than the cost (LCOE)—

Range of levelized cost and levelized avoided cost by case, 2022
2016 dollars per megawatthour

■ LCOE □ LACE

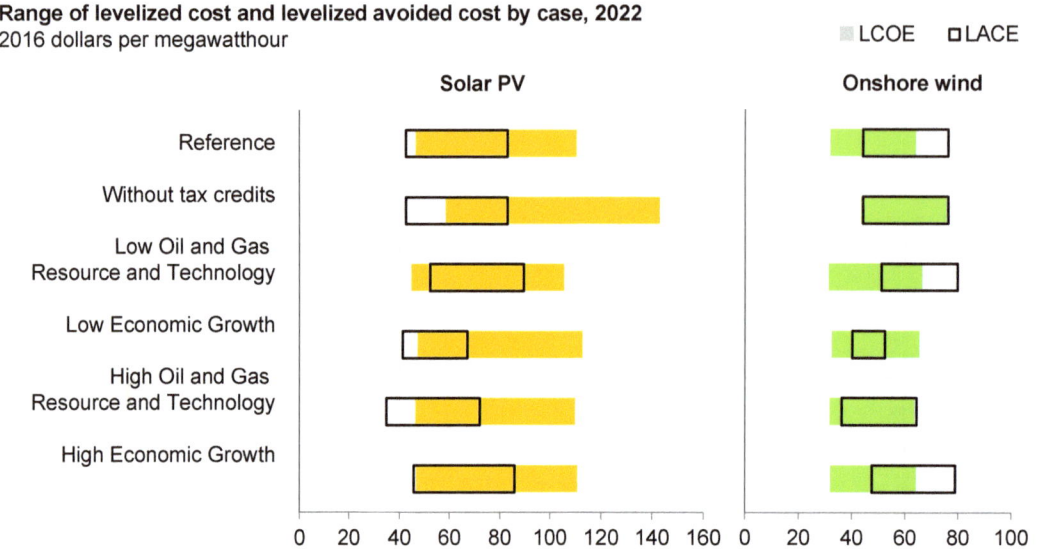

—particularly assumptions that affect natural gas price projections

- The availability of tax credits affects the effective cost of generation from solar and wind, but other policies may affect value.

- High or low natural gas prices, as respectively reflected in the Low and High Oil and Gas Resource and Technology cases, affect the cost of generation that wind or solar displaces, and thus play a big role in determining the value of these resources to the electric grid.

- Faster demand growth under high macroeconomic growth conditions increases the value of new generation resources. Slower macroeconomic growth, leads to relatively flat demand growth and less demand for new generation.

Transportation

Transportation energy consumption peaks in 2018 in the Reference case because rising fuel efficiency outweighs increases in total travel and freight movements throughout the projection period.

Transportation energy use declines between 2018 and 2034 in the Reference case—

Transportation sector consumption
quadrillion British thermal units

Transportation sector consumption
quadrillion British thermal units

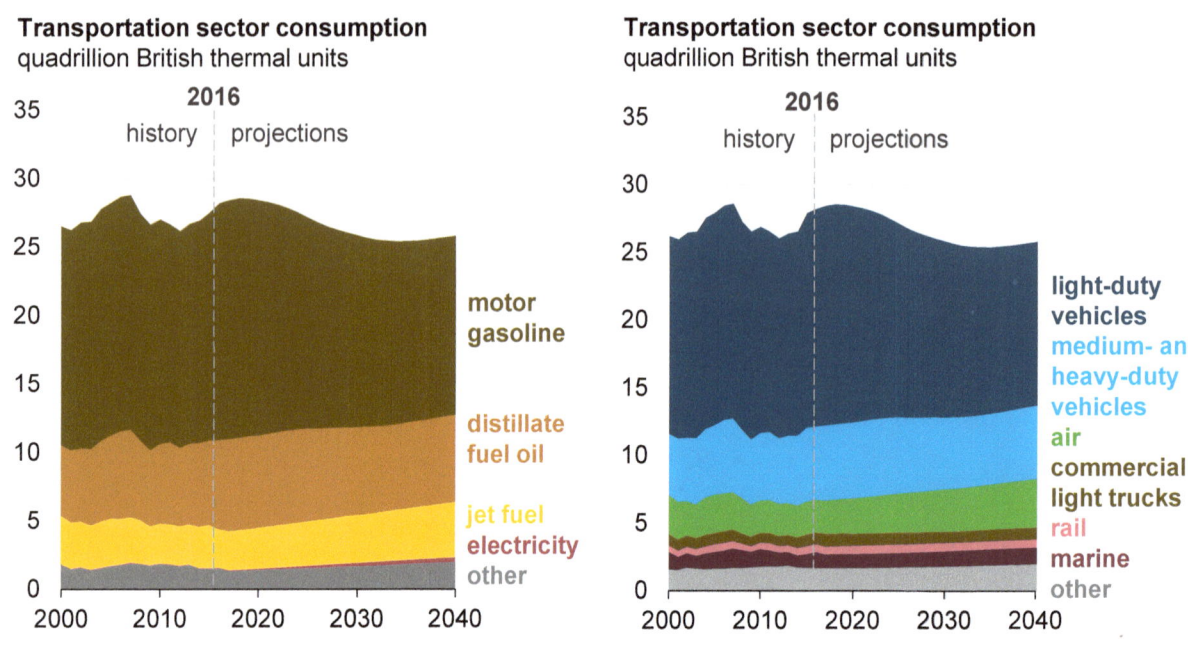

—driven by improvements in fuel economy

- Total transportation-related energy consumption peaks in 2018 in the Reference case and then declines through 2034 even as total travel and freight movement increases.

- Similarly, despite increases in light-duty travel, light-duty vehicle energy use also peaks in 2018 and then declines through 2040 as a result of higher fuel efficiency.

- Because the increase in freight travel demand is offset by rising fuel economy standards, heavy-duty vehicle energy consumption is approximately the same in 2040 as it was in 2016.

- Demand for air transport rises over the projection period, leading to an increase in energy used by air travel despite efficiency improvements.

Average light-duty fuel economy improves in the Reference case—

Light-duty stock fleet fuel economy
miles per gallon

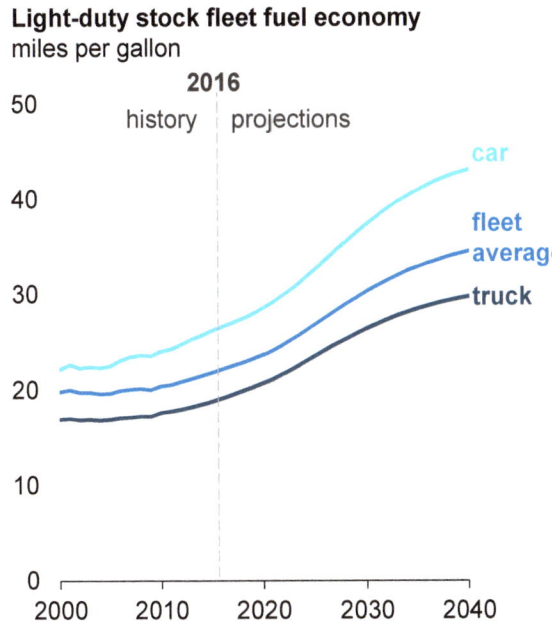

Light-duty vehicle sales shares
percent

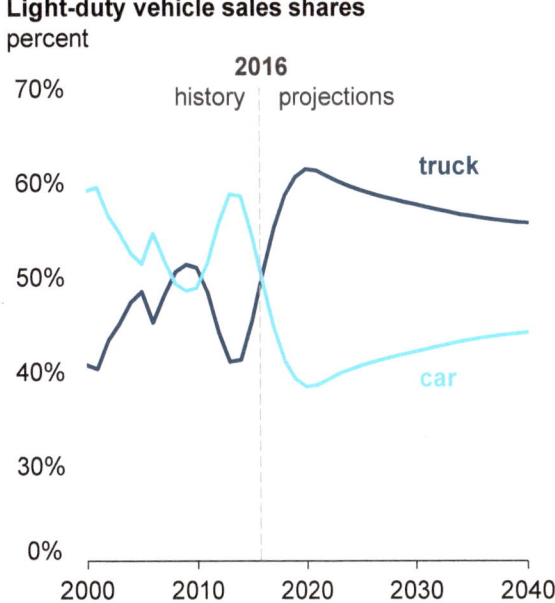

—even as the share of light-duty trucks increases

- Light-duty stock fuel economy is projected to rise from 22.2 miles per gallon (mpg) in 2016 to 34.6 mpg in 2040 in the Reference case. Current regulations require annual increases in fuel economy and reductions in greenhouse gas emissions through model year 2025, leading to a significant decrease in gasoline consumption.

- The sales share of light-duty trucks, which have lower fuel economy compared with passenger vehicles, limits the increase of the average fuel economy of the light-duty fleet.

- The shift toward light-duty trucks is driven by lower fuel costs and a changing preference for pickup trucks and sport utility vehicles rather than cars.

- Light-duty truck sales decrease after 2018 with the rise in popularity of front-wheel drive crossover vehicles that are classified as passenger cars.

With the second phase of fuel efficiency regulations, medium- and heavy-duty vehicle energy consumption declines over 2023–33—

Medium- and heavy-duty vehicle metrics

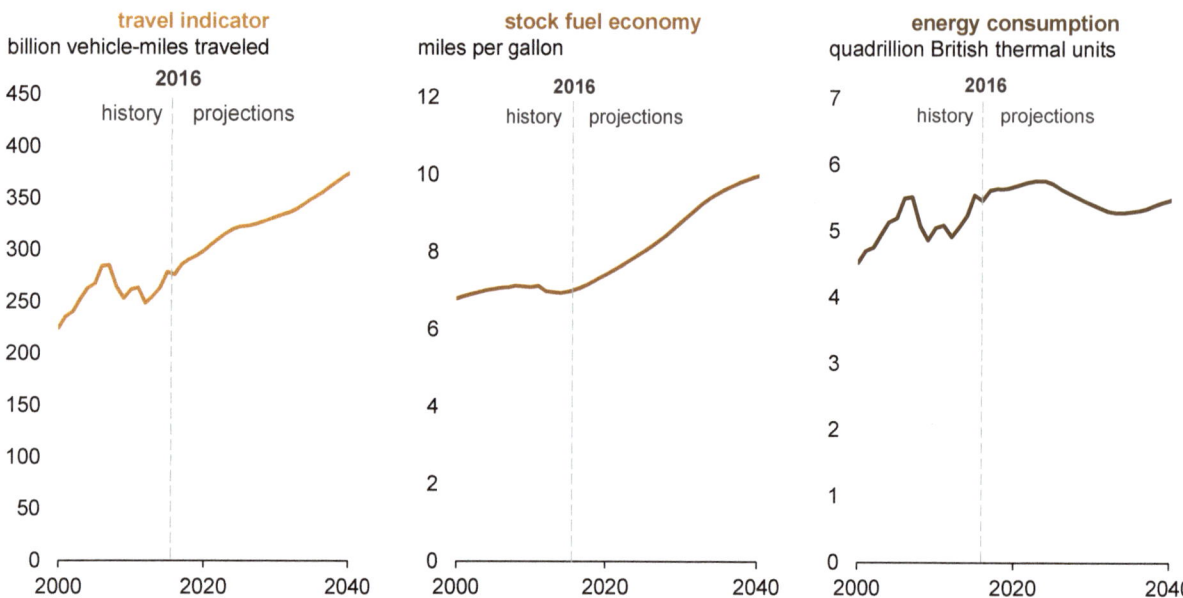

—despite continued increase in miles traveled

- The second phase of the fuel efficiency and greenhouse gas regulations for medium- and heavy-duty vehicles takes full effect in 2027.

- Fuel economy of new medium- and heavy-duty vehicles increases by 38% from 2016–32 before leveling off, but stock fuel economy continues to increase through 2040 as less fuel efficient vehicles retire.

- Energy consumption from medium- and heavy-duty vehicles decreases from 2023 through 2033 before increasing in the Reference case, where fuel economy standards for trucks do not increase beyond 2027.

- Diesel remains the dominant fuel for trucks despite increasing use of alternative fuels.

Sales of battery electric, plug-in electric hybrid, and fuel cell vehicles increase in the Reference case—

New light-duty vehicle sales
thousands of vehicles

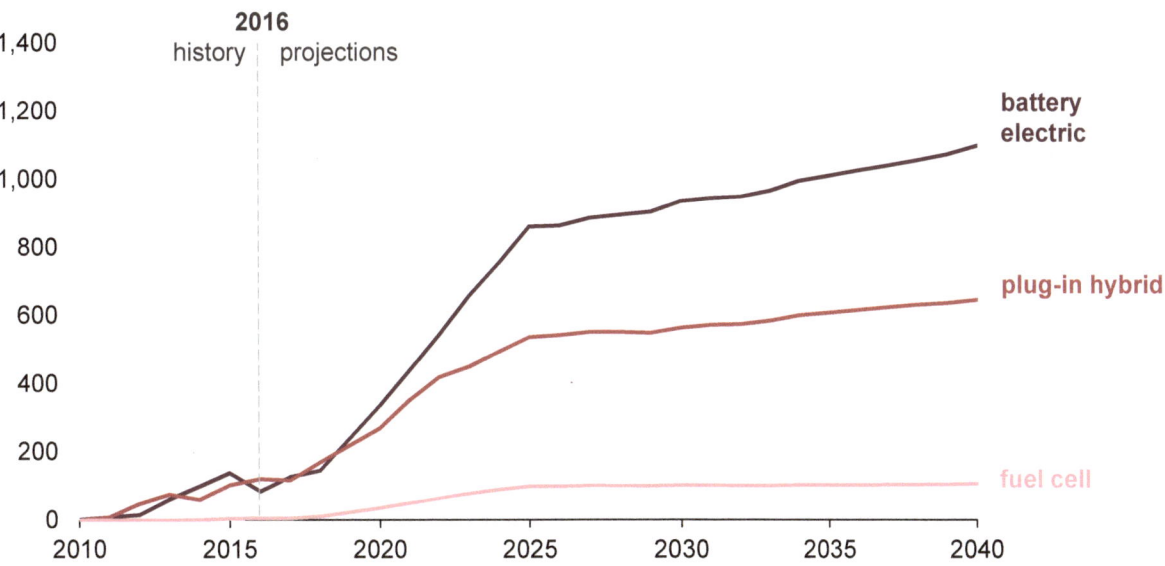

—because of lower projected battery costs and existing state policies

- Battery electric vehicles (BEV) sales increase from less than 1% to 6% of total light-duty vehicles sold in the United States over 2016–40, and plug-in hybrid electric vehicle (PHEV) sales increase from less than 1% to 4% over the same period. Hydrogen fuel cell vehicle (FCV) sales grow to approximately 0.6% of sales by 2040.

- In 2025, projected sales of light-duty battery electric, plug-in hybrid electric, and hydrogen fuel cell vehicles reach 1.5 million, about 9% of projected total sales of light-duty vehicles.

- Regional programs such as California's Zero-Emission Vehicle regulation, which has been adopted by nine additional states, and California's SB-32, which requires a reduction in greenhouse gas emissions, spur alternative vehicle sales, especially electric and fuel cell vehicles.

- Updated data that indicate lower battery costs have increased EIA's outlook for BEV and PHEV sales.

Even with improving commercial aircraft efficiency—

Air transportation metrics

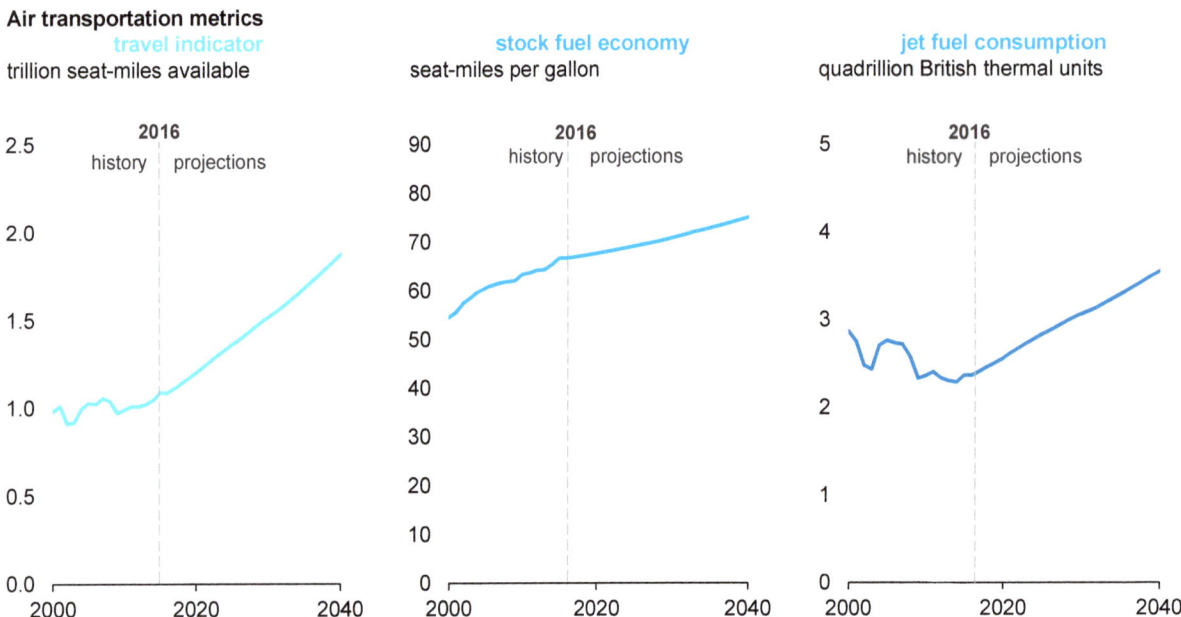

| travel indicator | stock fuel economy | jet fuel consumption |
| trillion seat-miles available | seat-miles per gallon | quadrillion British thermal units |

—jet fuel use rises in the Reference case with increased travel

- Jet fuel consumption increases more than 40% between 2016 and 2040 in the Reference case, as demand for air travel more than offsets projected efficiency gains in aircraft.

- With slow fleet turnover, aircraft stock efficiencies rise more than 12% between 2016 and 2040, as measured by seat-miles per gallon.

- U.S. load factors (fraction of filled seats and cargo space) for domestic and U.S. international routes, which increased significantly over 1995–2010, are projected to remain relatively flat over 2016-40.

- Even with the rise in aircraft efficiency, U.S. seat-miles more than double and freight revenue ton-miles nearly double through 2040, yielding a net increase in jet fuel consumption in the transportation sector.

Buildings

Despite growth in the number of households and the amount of commercial floorspace, improved equipment and efficiency standards contribute to residential and commercial consumption remaining relatively flat or declining slightly from 2016 to 2040 in the Reference case.

Residential and commercial fuel consumption are relatively stable in the Reference case—

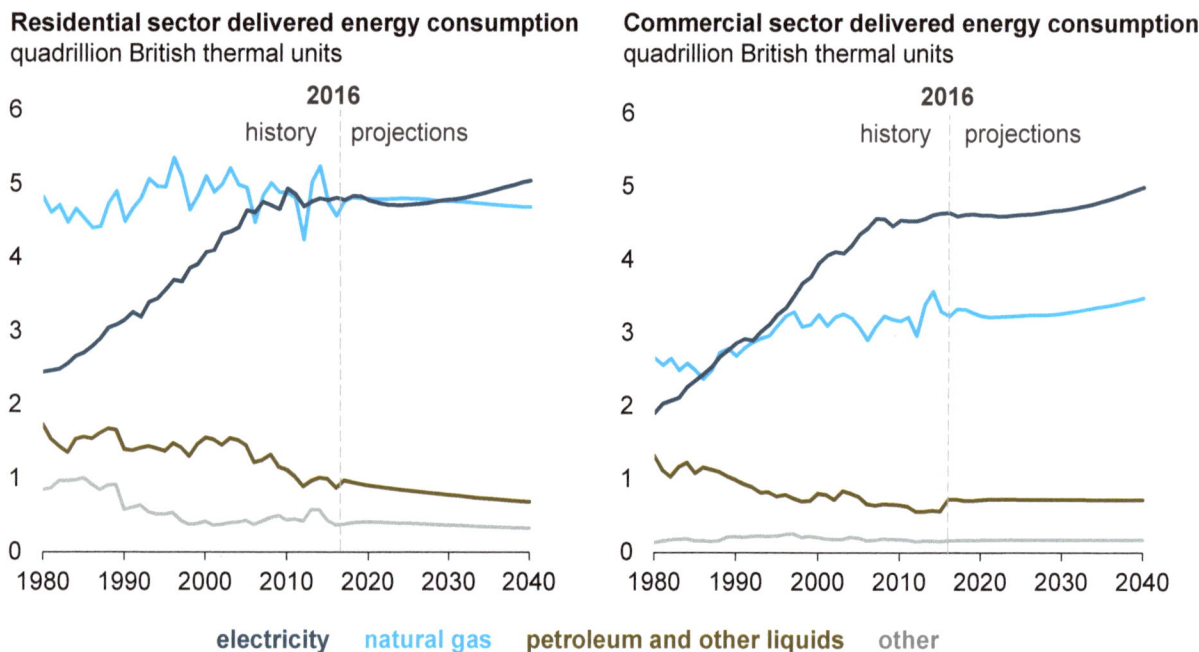

Residential sector delivered energy consumption
quadrillion British thermal units

Commercial sector delivered energy consumption
quadrillion British thermal units

electricity   natural gas   petroleum and other liquids   other

—as energy efficiency and other factors offset growth in end-use energy service demand

- Laws and regulations to introduce and update appliance standards and building codes have continued to increase energy efficiency in the residential and commercial sectors.

- Electricity demand in both sectors has been relatively flat in recent years, and it continues to be flat in the near term. Eventually, the increased adoption and saturation of new uses not currently covered by appliance standards increases consumption.

- Continued population shifts toward warmer parts of the country tend to lower heating demand and increase cooling demand. More energy is used for heating, so the result is a decrease in net delivered energy.

- Consumption of natural gas, used primarily for space heating, water heating, and cooking, has historically grown slower than electricity, and this trend generally continues through the projection.

- Use of petroleum-based fuels such as propane and heating oil continues to decline in the residential sector and remains relatively flat in the commercial sector.

Gradual increases in electricity and natural gas prices—

Electricity prices
2016 cents per kilowatthour

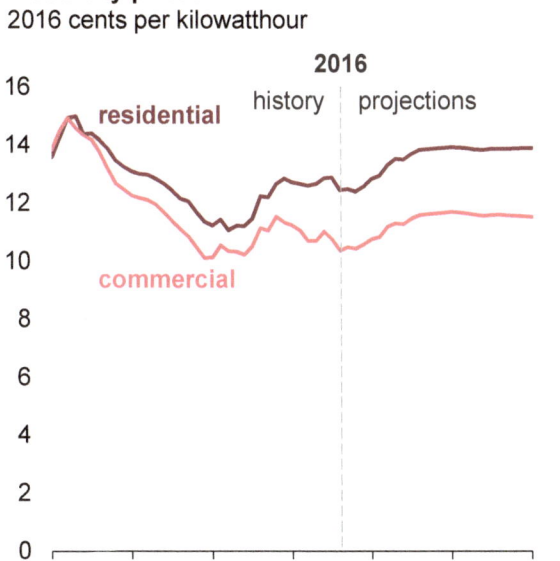

Natural gas prices
2016 dollars per thousand cubic feet

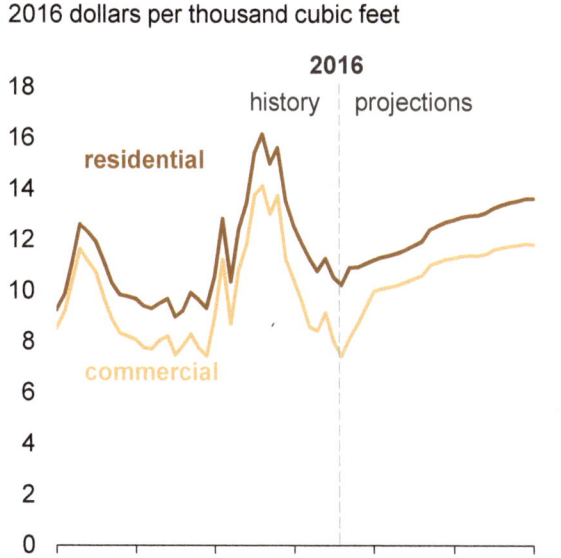

—affect residential and commercial energy consumption

- Following modest price increases from 2016 to 2030 in both residential and commercial sectors, electricity prices stabilize after 2030.

- As electricity prices flatten from 2030 to 2040, along with factors such as geographic population shifts and floorspace growth, electricity consumption rises at an increased rate in both sectors.

- Residential natural gas consumption is relatively stable, despite steadily increasing residential natural gas prices.

- Commercial natural gas prices increase in the near term, while commercial natural gas consumption remains flat; in the longer term, as price increases slow after 2030, commercial natural gas consumption begins to increase.

Energy consumption decreases for most major end uses in the residential and commercial sectors—

Residential sector delivered energy consumption
quadrillion British thermal units

Commercial sector delivered energy consumption
quadrillion British thermal units

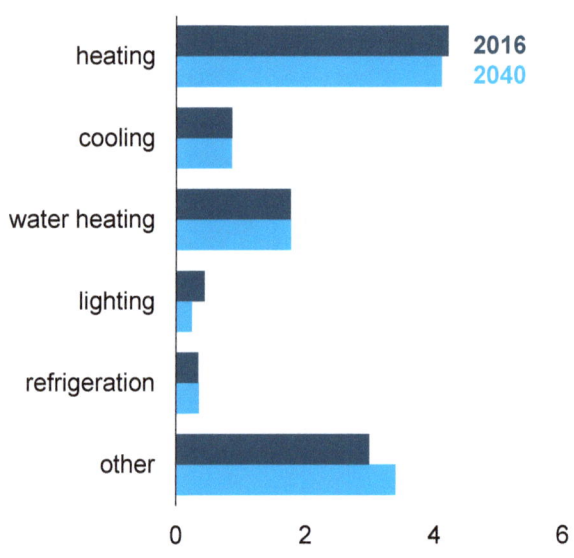

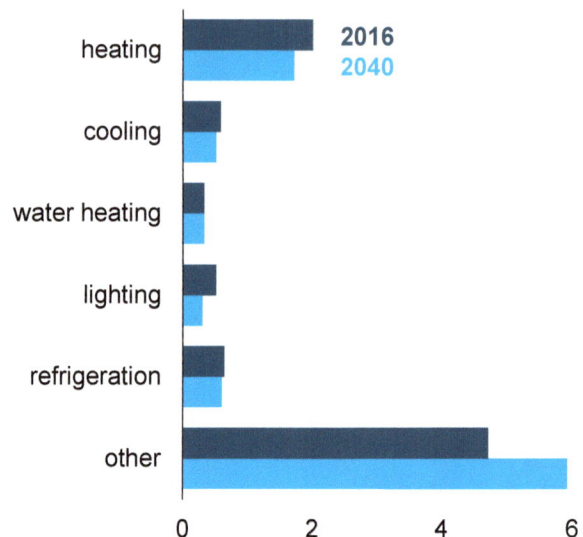

—with improved equipment efficiency and standards in the Reference case

- Energy consumption for lighting declines in the residential and commercial sectors as light-emitting diodes and compact fluorescent lamps continue to replace incandescent lamps and other bulb types.

- Energy consumption most residential and commercial applications either remains flat or declines slightly from 2016 to 2040 in the Reference case, despite growth in the number of households and the amount of commercial floorspace.

- Utility rebates contribute to a decrease in energy consumption. These rebates are expected to increase with the implementation of the Clean Power Plan (CPP) because energy efficiency programs are one of the available compliance strategies, and they are expected to grow more than they would in the absence of the CPP.

- In the residential sector, most of the growth in the *Other* category comes from increasing market penetration of smaller electric devices, most of which are not covered by efficiency standards.

- In the commercial sector, increased energy consumption for *Other* primarily reflects an increase in non-building uses such as telephone and technology networks.

Per-household electricity use continues to decline in the Reference case—

Residential electricity use per household
thousand kilowatthours per household

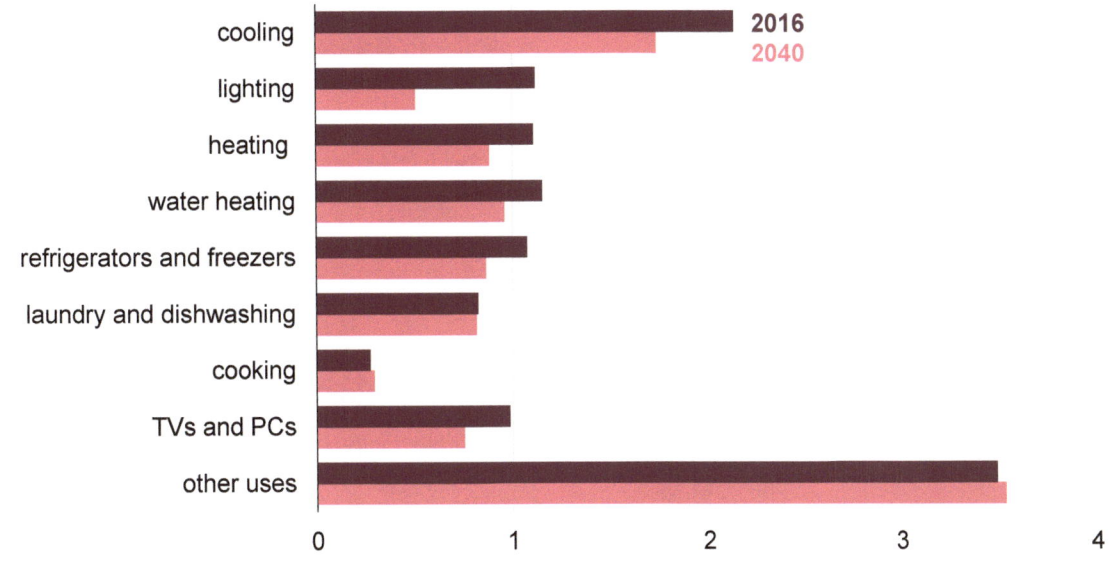

—led by efficiency improvements in lighting, cooling, and heating

- Electricity use per household continues to decrease in the Reference case, as household growth exceeds growth in residential electricity use.

- By 2040, the average household uses less than half as much electricity for lighting as they did in 2016, as customers replace incandescent bulbs with more energy efficient light-emitting diodes (LEDs) and compact fluorescent lamps (CFLs).

- Space cooling consumption for the average household declines by nearly 20%, as energy efficiency improvements more than offset the increased demand for space cooling.

- Per household electricity use by miscellaneous loads, a category that encompasses a wide range of equipment such as small electronic devices, home security systems, and pool pumps, increases slightly as efficiency improvements only partially offset the increased adoption and market penetration of new devices.

- Residential on-site electricity generation, mostly from photovoltaic solar panels, lowers total purchased delivered electricity from the electric grid.

AEO2017 includes new data from EIA's Commercial Buildings Energy Consumption Survey—

Commercial energy intensities, 2016
thousand British thermal units per square foot

Commercial floorspace by type, 2016
million square feet

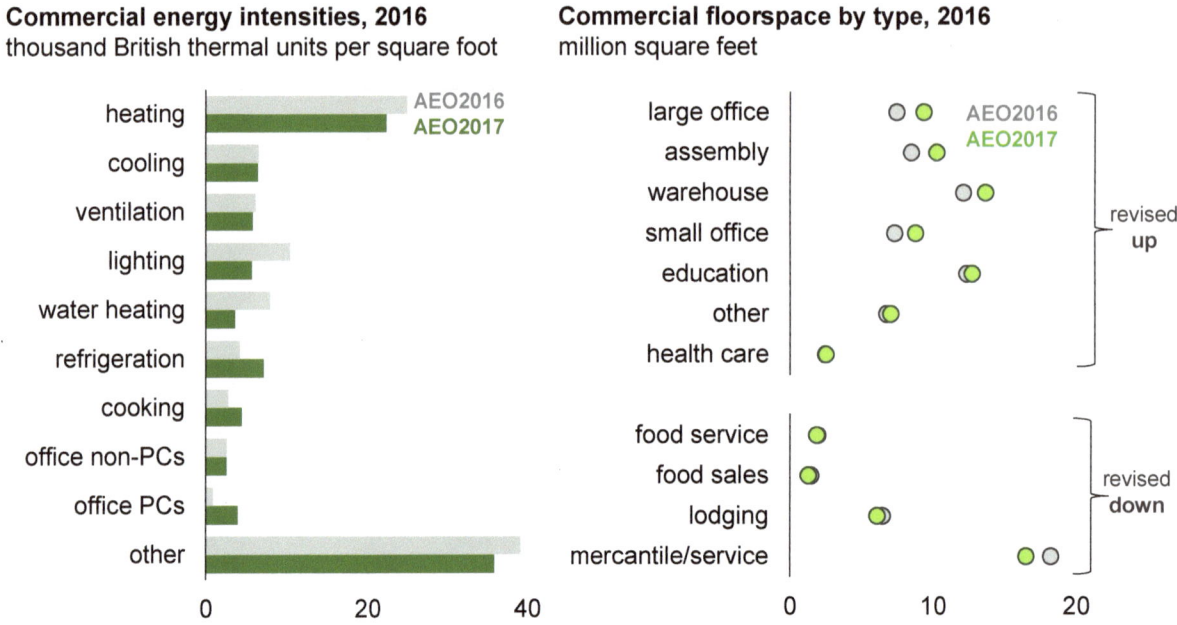

—leading to revisions in commercial building mix and energy consumption

- AEO2017 is based on the latest Commercial Buildings Energy Consumption Survey (CBECS), which was released during 2015 and 2016 and is the first update to be included in the AEO since AEO2007. The sample of buildings surveyed was drawn from the set of commercial buildings as of 2012.

- The latest CBECS provides a better understanding of the makeup of the commercial sector as well as the energy consumption associated with different end uses.

- Overall commercial floorspace is larger than previous estimates, especially for large offices and assembly buildings.

- Some end uses, particularly lighting and water heating, have changed significantly since the previous CBECS, which was based on the set of commercial buildings as of 2003 and did not consider as many building types as the latest CBECS.

- Categorization of some end uses in commercial buildings has changed. For instance, the category of office personal computers (PCs) now includes data center servers and all video screens; this equipment was previously categorized as *other end-uses*.

On-site electricity generation in residential and commercial buildings increases in the Reference case—

Buildings sector on-site electric generating capacity
gigawatts

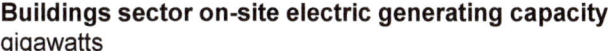

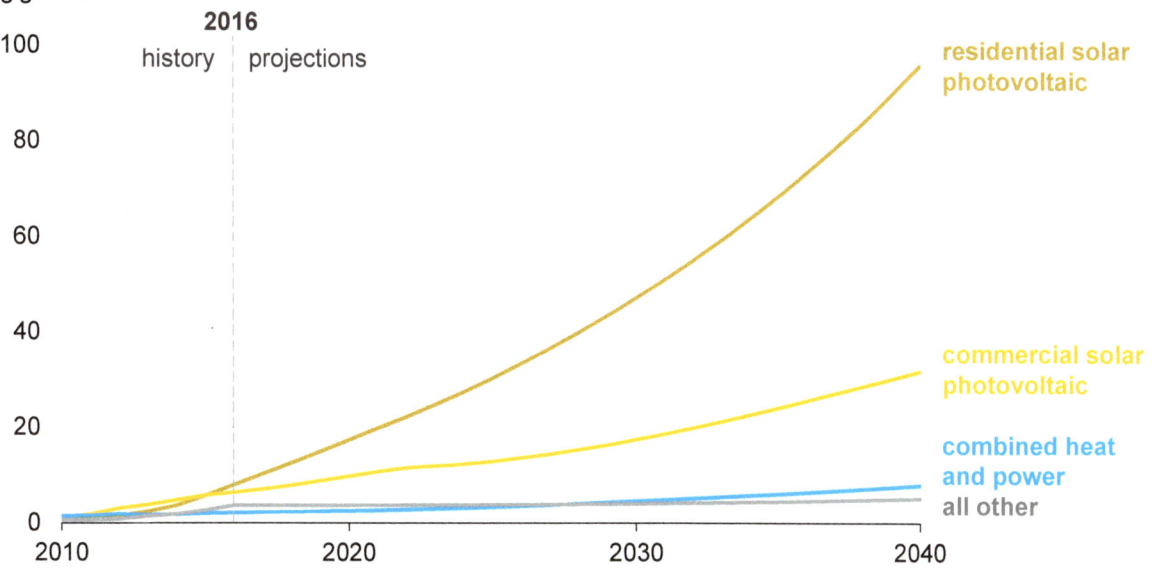

—reflecting declining technology costs and the continued availability of incentives for solar technologies to all sectors through 2021

- Solar photovoltaic (PV) systems account for most of the growth in buildings-sector on-site (or distributed) electricity generation in the AEO2017.

- Solar PV adoption grows from a 2010 base of less than 2 gigawatts (GW) in the residential and commercial sectors to more than 125 GW of capacity in 2040 in the Reference case.

- Other technologies such as small wind and combined heat and power, mostly in the commercial sector, grow more slowly and reach about 13 GW of capacity by 2040.

- Federal investment tax credits for solar technologies currently cover 30% of installed cost through 2019, dropping to 26% in 2020 and to 22% in 2021. In 2022, residential tax credits expire, and commercial credits are reduced to 10%.

- The differences from AEO2016 come from expected technology cost declines and changes in the way that EIA projects buildings will employ solar PV over time (adoption modeling). Additionally, EIA's new residential PV adoption projection uses econometric modeling of ZIP code-level solar resources, electricity rates, and financial metrics.

Industrial

With economic growth and relatively low energy prices, energy consumption in EIA's three industrial sub-sectors (energy-intensive manufacturing, non-energy-intensive manufacturing, and nonmanufacturing) increases during the projection period across all cases. Energy intensity declines across all cases as a result of technological improvements.

Industrial delivered energy consumption grows in all cases—

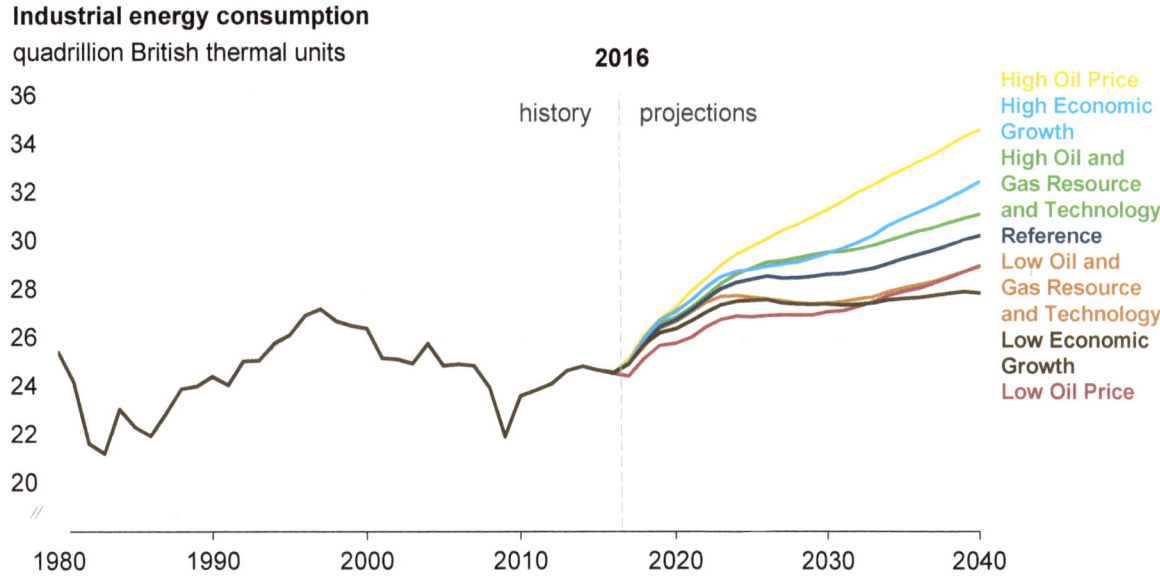

Industrial energy consumption
quadrillion British thermal units

—but is highest in the High Oil Price case and the High Economic Growth cases over most of the projection

- Reference case industrial energy consumption is projected to grow more than 25%, from 26 to 32 quadrillion British thermal units between 2016 and 2040.

- Industrial energy consumption is greatest in the High Oil Price case. Although industrial energy use grows in all cases, more energy is used to produce steel, fabricated metal products, and machinery in the High Oil Price case than the Reference case because of greater demand for these products.

- Combined heat and power (CHP) generation in the High Oil Price case is about 26%, or about 53 billion kilowatthours, above the Reference case by 2040 largely because of higher CHP generation for coal-to-liquids and gas-to-liquids. Coal-to-liquids and gas-to-liquids are economical in the High Oil Price case in the mid-2020s and after.

Industrial sector energy consumption grows faster than in other demand sectors in the Reference case—

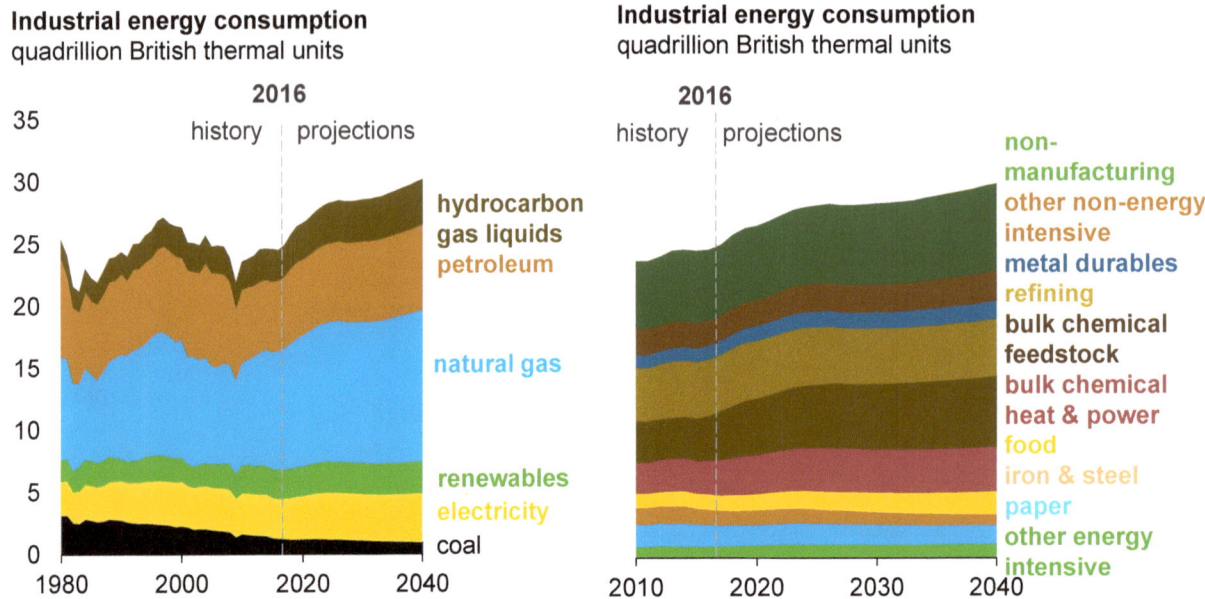

Industrial energy consumption
quadrillion British thermal units

2016
history | projections

hydrocarbon
gas liquids
petroleum

natural gas

renewables
electricity
coal

Industrial energy consumption
quadrillion British thermal units

2016
history | projections

non-manufacturing
other non-energy intensive
metal durables
refining
bulk chemical feedstock
bulk chemical heat & power
food
iron & steel
paper
other energy intensive

—led by increases in petroleum and natural gas consumption

- Driven by economic growth and supported by relatively low energy prices, industrial energy consumption in EIA's three main industrial sub-sectors (nonmanufacturing, energy-intensive manufacturing, and non-energy-intensive manufacturing) increases during the projection period across all cases.

- Natural gas (used for heat and power in many industries) and petroleum (a feedstock for bulk chemicals) make up the majority of delivered industrial energy consumption, followed by purchased electricity, renewables, and coal.

- Total industrial energy consumption growth averages nearly 1% per year from 2016–40 in the Reference case, the highest growth rate of any demand sector, as economic growth exceeds efficiency gains.

- Industrial coal usage declines by 24% over the projection period as its use in combined heat and power (CHP) is largely replaced by lower-cost natural gas.

- Hydrocarbon gas liquids (HGL) such as ethane, propane, and butane are largely produced by processing liquids from wet natural gas wells. HGL, which are widely used as feedstock in chemical processes, are a major source of growth in overall industrial use of petroleum.

Industrial energy intensity declines across all subsectors—

Industrial energy intensity (Reference case)
trillion British thermal units per billion dollars of shipments

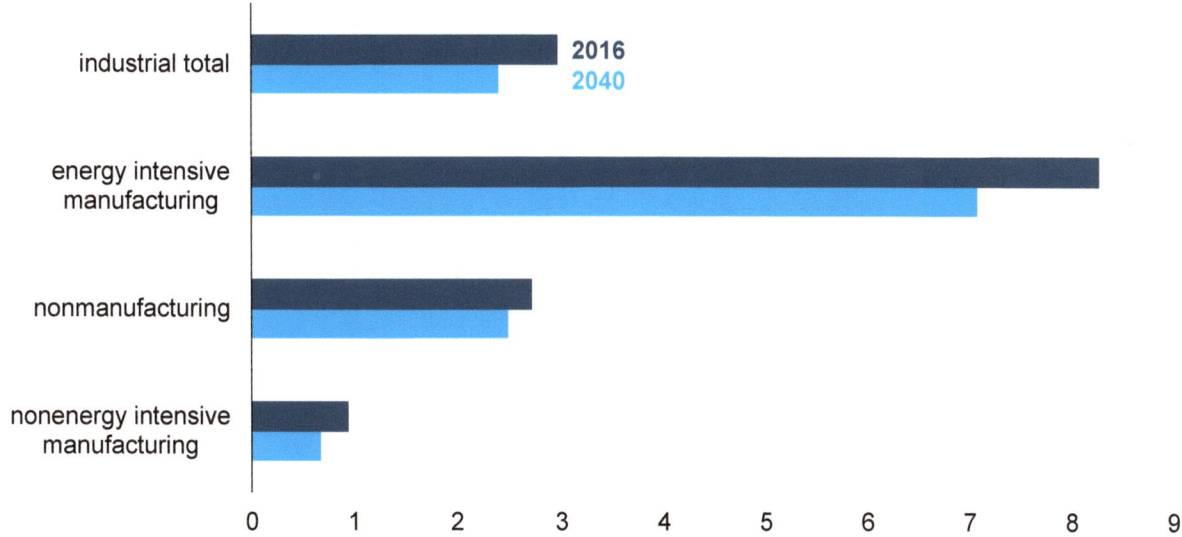

—moderating energy consumption increases

- Overall industrial energy intensity, measured as energy consumption per industrial shipment, declines by approximately 0.9% per year from 2016 to 2040 in the Reference case, consistent with historic trends.

- Manufacturing energy intensity declines as a result of continued efficiency gains in industrial equipment as well as a shift in the share of shipments from energy-intensive manufacturing industries to other industries.

- Energy-intensive industries, which include food, paper, bulk chemical, glass, cement, iron and steel, and aluminum products, dominate overall industrial energy use consumption, accounting for less than 25% of industrial shipments but more than 60% of industrial energy use.

Industrial combined heat and power use grows in the Reference case—

Combined heat and power output
billion kilowatthours

Combined heat and power output
billion kilowatthours

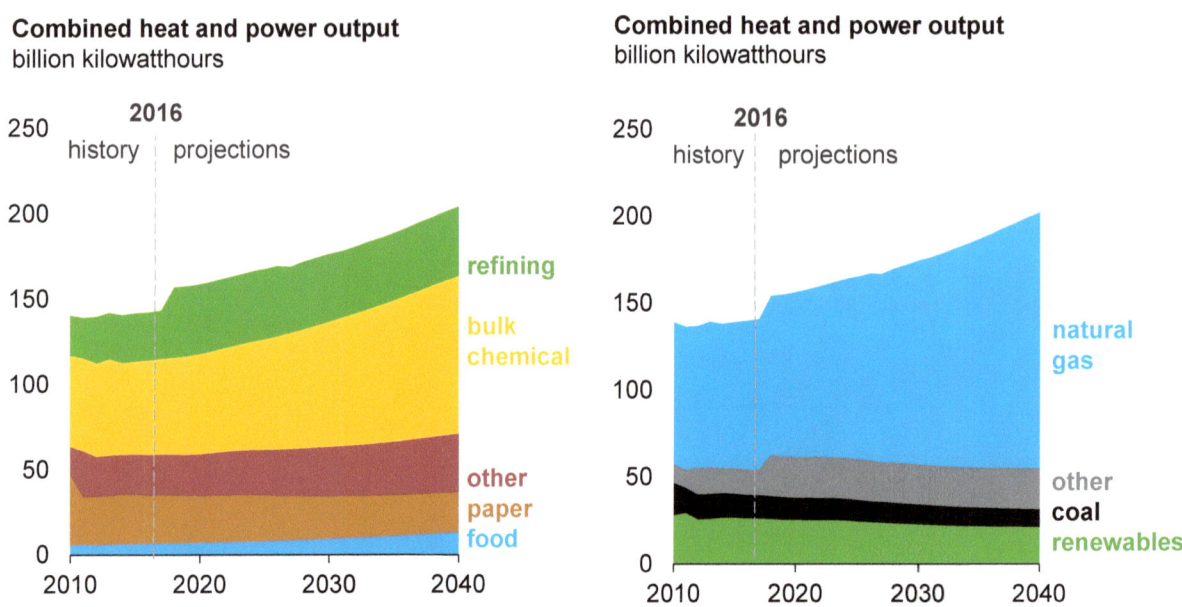

—as bulk chemicals and food are the fastest growing industries through 2040

- Natural gas is the most common fuel used in combined heat and power (CHP), but renewables are used in the paper industry. Specialty fuels such as blast furnace gas and still gas are used in the iron and steel industry and the refining industry, respectively.

- Industrial CHP is most commonly found in large, steam-intensive industries, such as bulk chemicals, refining, paper, and food.

- The median size of an industrial sector CHP facility is 30 megawatts (MW), and an average size of 65 MW. CHP offsets approximately 0.5 quadrillion British thermal units (Btu) of purchased electricity in 2016 and 0.7 quadrillion Btu in 2040.

References

Contacts

AEO Working Groups
https://www.eia.gov/outlooks/aeo/workinggroup/

AEO Analysis and Forecasting Experts
https://www.eia.gov/about/contact/forecasting.php#longterm

Topic	Subject matter expert contact information		
General questions	Angelina LaRose	202-586-6135	angelina.larose@eia.gov
Carbon dioxide emissions	Perry Lindstrom	202-586-0934	perry.lindstrom@eia.gov
Coal supply and prices	David Fritsch	202-287-6538	david.fritsch@eia.gov
Commercial demand	Kimberly Klaiman	202-586-1678	kimberly.klaiman@eia.gov
Economic activity	Vipin Arora	202-586-1048	vipin.arora@eia.gov
Electricity generation, capacity	Jeffrey Jones	202-586-2038	jeffrey.jones@eia.gov
Electricity generation, emissions	Laura Martin	202-586-1494	laura.martin@eia.gov
Electricity prices	Lori Aniti	202-586-2867	lori.aniti@eia.gov
Ethanol and biodiesel	Sean Hill	202-586-4247	sean.hill@eia.gov
Industrial demand	Kelly Perl	202-586-1743	eia-oeceaindustrialteam@eia.gov
International oil demand	Linda Doman	202-586-1041	linda.doman@eia.gov
International oil production	Laura Singer	202-586-4787	laura.singer@eia.gov
National Energy Modeling System	Daniel Skelly	202-586-1722	daniel.skelly@eia.gov
Nuclear energy	Michael Scott	202-586-0253	michael.scott@eia.gov
Oil and natural gas production	Terry Yen	202-586-6185	terry.yen@eia.gov
Oil refining and markets	William Brown	202-586-8181	william.brown@eia.gov
Renewable energy	Christopher Namovicz	202-586-7120	christopher.namovicz@eia.gov
Residential demand	Kevin Jarzomski	202-586-3208	kevin.jarzomski@eia.gov
Transportation demand	John Maples	202-586-1757	john.maples@eia.gov
Wholesale natural gas markets	Kathryn Dyl	202-287-5862	kathryn.dyl@eia.gov
World oil prices	Laura Singer	202-586-4787	laura.singer@eia.gov

For more information

U.S. Energy Information Administration homepage | www.eia.gov

Short-Term Energy Outlook | www.eia.gov/steo

Annual Energy Outlook | www.eia.gov/aeo

International Energy Outlook | www.eia.gov/ieo

Monthly Energy Review | www.eia.gov/mer

Today in Energy | www.eia.gov/todayinenergy

www.ingramcontent.com/pod-product-compliance
Lightning Source LLC
Chambersburg PA
CBHW051045180526
45172CB00002B/523